BIOCHEMISTRY

BIOCHEMISTRY

Editor
N.P. SHARMA

KALPAZ

Published by

Kalpaz Publications
C-30, Satyawati Nagar,
Delhi – 110052
Ph.: 011-47034999, 9811692060
www.kalpazpublications.com
E-mail: kalpaz@hotmail.com

Distribution Network
gyanbooks.com
India, USA, Canada, UK, Australia, France

ISBN: 978-93-5722-941-8 (PB)

First Published, 2024

Printed at: Gyan Press, Delhi.

Biochemistry
Author: N.P. Sharma

Preface

Biochemistry is the branch of science which deals with the chemical processes that go on in living matter, ranging from viruses and bacteria to plants and animals. Consideration of this concept leads to a functional definition of biochemistry as the science concerned with the chemical constituents of living cells and with the reactions and processes they undergo. By this definition, biochemistry encompasses wide areas of cell biology and molecular biology. Actually, it is a study of biology at molecular level. It is concerned with the molecules that make up the structure of cells and organs, i.e., molecular anatomy.

As biochemistry deals with carbon compounds and the reactions they undergo in living organism, the chemistry of carbon compounds—organic chemistry—is found to be basic to biochemistry. The chemistry of natural products and biochemistry overlap extensively and lack a clearly defined border. Biochemistry is also concerned with molecular physiology, i.e., the functions of molecules in carrying out the needs of the cells and organs.

A field of biochemistry, that has stimulated lively interest, is biochemical genetics. Perhaps the most exciting area of preset research in biochemistry deals with mechanism for the regulation of the synthesis of cellular compounds. Broadly speaking, biochemistry may be treated as a discipline in which biological phenomena analysed in terms of chemistry. Due to this reason, the biochemistry is also named as biological chemistry or chemical biology.

As the knowledge of biochemistry is increasing rapidly, newer disciplines such as enzymology (science of the study of enzymes), endocrinology (science of the study of endocrine secretions or the hormones), clinical biochemistry, molecular biochemistry, etc., are emerging from the parent biochemistry. Further, agricultural biochemistry, pharmacological biochemistry, etc., have also come up as certain link specialities.

The present book provides an up-to-date account of biochemistry. Students and teachers from different branches of sciences will find this work highly informative.

Editor

Contents

Chapter 1

Introduction

Biochemistry may be defined as that branch of science which deals with the chemical processes that go on in living matter, ranging from viruses and bacteria to plants and animals. Biochemistry can be defined more formally as the science concerned with the chemical basis of life (Gk *bios* 'life'). The cell is the structural unit of living systems. Consideration of this concept leads to a functional definition of biochemistry as the science concerned with the chemical constituents of living cells and with the reactions and processes they undergo. By this definition, biochemistry encompasses wide areas of *cell biology* and all of *molecular biology*. Actually, it is a study of biology at molecular level. It is concerned with the molecules that make up the structure of cells and organs, i.e., molecular anatomy.

The term Biochemistry *(Bios* = life) was first of all introduced by a German Chemist Carl Neuberg in 1903. In terms of history, biochemistry is considered to be a young science. It started largely as an offshoot of organic chemistry and later incorporated ideas obtained from physical chemistry. Theophrastus Bombastus von Hohenheim (1491-1641), better known as Paracelsus, was the earliest to refer to some chemical aspects of medicine in his writings. First of all, he attained the knowledge of chemistry of his time arid then entered the field of medicine to apply his knowledge of chemistry. Later, his followers especially *Jan Baptist van Helmont* (1577-1644) amalgamated 'he science of chemistry with medicine which appeared under the name of 'medical chemistry' (or latrochemistry).

The chemistry of natural products and biochemistry overlap extensively and lack a clearly defined border. Biochemistry is also concerned with molecular physiology, i.e., the functions of molecules in carrying out the needs of the cells and organs. Again, there is a considerable information in this area and the knowledge about the chemical reactions taking place in cells is increasing rapidly. A field of biochemistry, that has stimulated lively interest, is biochemical genetics.

Perhaps the most exciting area of preset research in biochemistry deals with mechanism for the regulation of the synthesis of cellular compounds. Broadly speaking, biochemistry may be treated as a discipline in which biological phenomena analysed in terms of chemistry. Due to this reason, the biochemistry is also named as *biological chemistry or chemical biology.*

Modern biochemistry has two separate branches:

—Descriptive biochemistry and
—Dynamic biochemistry

Descriptive biochemistry deals with the qualitative and quantitative characterisation of the various cell components. This branch is more a concern of the organic chemist.

Dynamic biochemistry deals with the elucidation of the nature and the mechanism of the reactions involving the cell components. This branch of biochemistry has now become the language of modern biochemistry. As the knowledge of biochemistry is increasing rapidly, newer disciplines such as *enzymology* (science of the study of enzymes), *endocrinology* (science of the study of endocrine secretions or the hormones), *clinical biochemistry, molecular biochemistry,* etc., are emerging from the parent biochemistry. Further, *agricultural biochemistry, pharmacological biohemistry,* etc., have also come up as certain link specialities.

Biochemistry shows that living organisms are composed of similar organic *compounds, in particular the* proteins and nucleic acids. These could have been synthesised in conditions prevailing over 4000 million years ago on earth and by a process, of chemical evolution into a DNA nucleic acid substance itself. The achievements of biochemistry and the activities of biochemists can be divided, into several large categories.

1. *Kinds and structure of organic compounds:* Fundamental to all other activity has been the compilation of a list of the organic compounds to be found in living organisms together with the elucidation of the structures of each. These myriads of compounds range from the simpler amino acids, sugars, and fatty acids, to the pigments responsible for the colour of flowers, the vitamins and coenzymes, and large macromolecular polymers such as proteins and nucleic acids.

2. *Metabolism and Biosynthesis:* The activity that has, perhaps, been most spectacularly successful has been the elucidation of the metabolic pathways for the biosynthesis of naturally occurring compounds from their precursors, that is, from foodstuffs in the case of animals or from carbon dioxide and minerals in the case of plants through photosynthesis. Biochemists have sue much, cif the details of the major metabolic, pathways for the synthesis and degradation of naturally occurring organic compounds in animals, plants, and microorganisms such as bacteria.

3. *Structure and Biological Function of Macromolecules* : A third area of interest relates to a correlation between the structure of macromolecules and their biological function. Thus, biochemists seek to learn how specific proteins serve as catalysts for specific chemical reactions; how the complex polysaccharides of cell walls and membranes serve the cells ;it which they are found; and how the complex lipids of the nervous system participate in the functions carried out by nervous cells,

4. *Cell Function:* Yet another aspect of biochemistry is the search for understanding the way in which the specialized cells of an animal or plan, conduct their unique functions such as how muscle cells contract, how cells make bone, and how red blood cells transport oxygen and carbox dioxide in the body, and how flowers make their pigments.

5. *Relations to Genetics:* Since in 1940s, studies of fungi and bacteria, and later of higher organisms, including man, have shown that genetic mutations often result in the loss of ability to carry out some specific biochemical reaction These observations led to the formation of the current concept of a gene as the in a chromosome that directs the biosynthesis of a specific protein. When that protein is an enzyme and the gene has been altered, as in a mutation, the cell loses the ability to conduct the reaction for which the enzyme is normally responsible.

The genes have row beer identified as specific portions of molecules of deoxyribonucleic acid (DNA), which can replicate themselves and direct the synthesis of specific protein within the cell. Much of the current work in biochemistry now involves the study of the details of nucleic acid and protein synthesis and is closely related to genetics. Indeed, the two sciences together are often referred to as molecular biology.

6. *Human Disease:* No biochemical research is more important than the continuing effort to understand human disease. Every year, increasing numbers of disorders are understood in terms of aberrations from normal pathways of metabolism. Biochemists and physicians have combined to unravel the nature of the fundamental disturbance in such diseases as diabetes and sickle cell disease. More than 800 such metabolic disturbances have been ,raced to hereditary gene defects, and alleviations for some of these conditions have been, found.

Table 1. The major causes of diseases.

1. *Physical agents:* Mechanical trauma, extremes of temperature, sudden changes in atmospheric pressure, radiation, electric shock.
2. *Chemical agents and drugs:* Certain toxic compounds, therapeutic drugs, etc.
3. *Biologic agents and drugs:* Viruses, rickettsiae, bacteria, fungi, higher forms of parasites.
4. *Oxygen lack:* Loss of blood supply, depletion of the oxygen-carrying capacity of the blood, poisoning of the oxidative enzymes.
5. *Genetic:* Congenital, molecular.
6. *Immunologic reactions:* Anaphylaxis, autoimmune disease.
7. *Nutritional imbalances:* Nutritional deficiencies, nutritional excesses.
8. *Endocrine imbalances:* Hormonal deficiencies, hormonal excesses.

Studies of nongenetic conditions have also been important. Investigation of the salt and acid base composition of blood plasma in health and disease have made possible the practice of major surgery without danger of shock or as well as the successful therapy of pernicious vomiting, infantile and many other disorders.n All diseases ate manifestations of abnormalities of molecules, chemical reactions, or processes. The major factors responsible for causing diseases in animals and humans are listed Table 1. All of them affect one or more critical chemical reactions or molecules in the body.

Importance of Biochemistry

Whenever a student studies biochemistry, he must have an understanding of the following facts:

i) *The structures and properties of substances constituting the frame work of cells and tissues:* The pioneers of biochemistry showed the presence and nature of substances called proteins, nucleic acids, polysaccharides, and complex lipids. The rate of progress of biochemical understanding has been paced by the development of procedures to isolate and purify such materials. New physiochemical methods showed molecular weights to range from 10,000 to more than 100,000,000 for individual substances. For years, the difficult task of establishing the complete structures of such molecules had been found to be experimentally unapproachable. By using, improved analytical techniques and a variety of new physical instruments, including degradative procedures that take advantage of the known specificities of hydrolytic enzymes, the ultracentrifuge, electrophoresis, recording spectrophotometers, spectropolarimeters, amino acid analyzers and X-ray crystallographic analysis, it becomes possible to know the general structures of these molecules.

   Detailed three-dimensional models of several smaller proteins and nucleic acids have been given and there has been rapidly growing understanding of the forces that are able to maintain these molecules, which would otherwise be long, thin, fibber-like structures, folded upon themselves into highly specific compact structures. The biological functioning of these molecules has been found to depend mainly upon their three-dimensional structures.

Understanding of the structures of these large molecules has been rapidly expanding, thereby offering the basis for a more penetrating insight into the mode of operation of enzymes, the structural basis for genetic phenomena, and fine structure of living cells.

(ii) *The structures and properties of substances which enter the cell as useful working materials or sources of energy or leaving the cell as waste products.*

iii) *The catalytic tasks of enzymes:* The study of biochemistry helps us to know bow the enzymes accomplish their catalytic tasks. Hundreds of enzymes, each more or less specific for one chemical reaction, have been isolated and many crystallized; each has proved to be 4 distinct, unique protein. This fascinating aspect of science, in many respects forms the heart of biochemistry.

iv) *The chemical processes which convert diet into the compounds which are characteristic of the cells of a given species.* Biochemistry has helped us to know the chemical processes which are used to convert the materials of the diet into the compounds which are characteristic of the cells of a given species. Study of the manifold individual events of metabolism was the centre of biochemical interest.

v) *Using the potential energy obtained from the oxidation of food stuffs utilised to drive the manifold energy-requiring process of the living cell:* The study of biochemistry helps us to know the synthesis of hundreds of new molecular species, intracellular accumulation of inorganic ions and organic compounds against concentration gradients and the performance of mechanical work. The study of biochemistry also helps us to know the impossibility of utilizing thermal energy to accomplish useful work at constant temperature making untenable a simple analogy between food-burning animals and fuel-burning heat engines, By understanding the biological solution to this problem, coupling of the oxidation of carbohydrates and fats to the synthesis of one compound, adenosine triphosphate (ATP), with subsequent utilization of the energy of this compound for virtually all endergonic processes, it has been found to be cardinal to the understanding of living cells.

vi) *Chemistry of inheritance:* Biochemistry helps us to know that it has been the presence, and activities of its complements of proteins that ascertain the form, organisation, and functions of a cell, i.e., its life. It is evident that the genetic "instructions" to a cell must offer the information needed to achieve= the precise synthesis of the ensemble of protein characteristic of that cell. This information gets encoded in the structure of the very large molecules of deoxyribonucleic acid (DNA). Cell duplication needs perfect reproductions of these molecules with subsequent equal distribution of the information between the cells. Utilization of this information needs its transmission from nucleus to the ribosome protein factories. Changes in the chemical structure of DNA got evident as mutations observed in subsequent generations. The totality of this information is able to make intelligible the laws of genetics, the nature and basis of hereditary diseases, and the biochemical operation of the process of evolution

vii) *The molecular basis of life*: The study of biochemistry helps us to study the molecular basis of life. Some of the main achievements of biochemistry are the discovery of the structure of DNA, the determination of the three dimensional structure of some protein molecules, and the unfolding of the central metabolic pathways. Organisms as diverse as the bacteria and mammals possess many common features at the molecular level. They employ the similar building units to construct macromolecules, The flow of genetic information from DNA through RNA to protein has been reported to he essentially the same in both types of organisms.

(viii) *In diagnosing:* The study of biochemistry plays a vital role in clinical diagnosis. For example, the level of certain enzymes in the scrum reveals many diseases.

(ix) *Discovery of new drugs:* The study of biochemistry has helped to provide a basis for the specific design of drugs.

(x) *Solving some fundamental problems in biology and medicine:* The rapid development of biochemistry in recent years has helped scientists involving some of the most challenging and fundamental problems in biology and medicine, For example, how is a single cell able to give rise to cells as different as those in muscle, the kidney and, the brain? How are cells able to find each other in forming a complex organ. What forms the mechanism of memory? How does light bring about a nerve impulse in the retina of eye ? What have been the causes of Schizophrenia? What have been the causes of cancer? Since the time of Aristotle, students of biology try to correlate structure and function, and this endeavour continues. The correlation of biological function and molecular structure forms the main theme of biochemistry.

From the above it follows that the main objective of biochemistry is to fill the wide gap between the highly integrated functions of the living cells and the various properties of its individual chemical constituents. Thus, *a biochemist is an investigator who utilises chemical, physical or biological techniques to study the chemical nature and behaviour of living matter.*

History of the development of biochemistry

The basis of biochemistry was, in fact, laid down by chemists like Scheele and Lavoisier. Early chemists were mostly busy in studying the chemistry of plant and animal materials. The Swedish pharmacist *Seheele,* (1742-1786) isolated and characterised many substances such as critic acid, lactic acid and uric acid and from biological materials.

The classic research done by *Lavoisier* (1743-1794) on oxidation and the role played by the oxygen in the process led him to investigate "burning" in the body and he came to the conclusion that oxygen is consumed in the reaction, the carbon dioxide is eliminated, and that heat is evolved. This may be said to be the beginning of modern 'animal calorimetry.' Lavoisier is often spoken of as *'father of modern biochemistry.' Wohler* (1800-1802) synthesised urea which is the end product of nitrogenous materials in the body. This was an achievement that did much to destroy the notion that animal products were endowed with a vitalism which made them fundamentally different from lifeless substances.

Experiments of Wohler was followed by the synthesis of acetic acid by Adolf Kolbe in 1884 and the synthesis of several organic compounds by Marcellin Berthellot in 1850s. Vitalism was quietly laid to rest. Organic synthesis remains very much alive. *Jestus von Liebig* (1803-1873) has been named as the *'father of agricultural chemistry'*. He showed that "the nutritive materials of all greet plants are inorganic substances." He authored many books which offered an impetus in the early development of chemistry. His most interesting book is *'Organic Chemistry in its Applications to Physiology and Pathology,'* which was published in 1842.

Although nucleic acids have been the newest of the four great groups of biochemical materials, their discovery dates back to observations by *Friedrich Miescher* in 1871. His discovery of nucleic acids in the nuclei of pus cells, obtained from discarded surgical bandages, made him to investigate the distribution and the properties of these compounds.

During the first half of the nineteenth century, the studies on heat conducted mainly by *Meyer and von Helmholtz* gave rise to the formulation of the 'laws of thermodynamics' which have been essential to the understanding of energy relations in biological systems. Besides respiration, the other physiological process which has attracted the attention of biochemists had been that of digestion. Main contributions towards this had been made by *van Helmont, Lazaro Spallanzani* (1729-1799), *Rene de Reaumur, William Benumont and Claude Bernard.* Claude Bernard (1813-1878) of Paris was considered to be the greatest of these. His contributions included the discovery of liver glycogen and its relation to blood sugar in health and disease. The process of fermentation has been probably the single most important process around which the interest of biochemists lasted fora considerable period, by about 1780, fermentation was recognized by *Theodor Schwann* (1810-1882) as a biological process. He ascertained that yeast had been a plant capable of converting sugar to ethanol and carbon dioxide. However, many of the leading chemists of the day, including Berzelius, Wohler and Leibing, regarded yeast to be nonliving and fermentation to be caused mainly by oxygen. Ridicule by Lie bing and Wohler delayed acceptance of Schewann's views until the illustrious French microbiologist *Louis Pasteur* (1822-1895) gave evidence overwhelming all objections. He founded the useful branch of Microbiology in 1857 and was successful in identifying several organisms that carried out various fermentation, including that leading to butyric acid, a type performed by organisms that function without oxygen. He gave definition of fermentation as "*la* vie sons V air" (life without air).

Pasteur, thus, gave the concept of aerobic and anaerobic organisms and their associated fermentations. These conclusions again aroused the ire of Liebing who again took sharp exception but this time with less effect. Such studies on fermentations had been climaxed by the demonstration in 1887 of *Eduard Buchner* (1860-1917) that sugars could be fermented by cell-free extracts of yeast. This discovery is considered to be the cornerstone of much of the enzymological and metabolic study of the twentieth century because it gave rise techniques of isolation and identification that ultimately allowed the study of enzymes and the individual reactions concerned. It also eliminated, yet once and for all, any traces of vitalism (or vitalistic theory) still lingering. The extensive researches carried out by *Pasteur* (1822-1895) led *Buchner* (1860-1917) to give modern concept of enzymes. It is to *Michaelis* we owe the present theory of enzyme-substrate interactions.

The concepts developed by *Arrhenius, van't Hoff* and *Ostwald* about osmatic pressure and electrolytes and the concept of pH developed by *Sorensen* were proved to be quite useful in understanding the functioning and homeostasis of the body fluids. Later many instruments like Van slyke blood gas apparatus, the ultracentrifuge of Svedber, and the electrophoresis apparatus of Tiselius were discovered.

The use of isotopes in biochemical research by *Urey* and *Schoenheimer* Synge, started a new era in modern biochemistry. Modern biochemistry may be said to have started with the demonstration by *Meyerhof* and *Hill* that there exists a correlation between lactic acid production in contracting muscle and oxygen consumption and heat evolution. The excellent research work carried out by *Warburg, Wieland, Keilin and Theorell* led to the discovery of enzymes and cofactors which are involved in cellular oxidation.

Lipmann's research led to the concept that the radiant energy of the sun utilised by the green plants for the synthesis of foodstuffs is converted in the animal cell into terminal pyrophosphate groups of ATP during the intermediate steps of oxidation. The significance of unknown food factors was understood by *Frederick Gowland Hopkins;* at Cambridge University and his associates, who gave the concept of deficiency diseases. Extensive series of feeding experiments utilizing synthetic diets had been carried out mainly by Babcook McCollum, Osborne, Mendel and Sherman. The research work done by Szent Gyorgyi and Kreb led to the discovery of sequence of reactions in the fate of lactate during aerobic oxidation. This was termed as Krebs or tricarboxylic acid cycle. It was also proved that the enzymes and cofactors were located in the mitochondrial particles of the cell.

Later studies revealed that the fatty acid and amino

acids, upon oxidation, also yield intermediates that have been identical with those in the Krebs cycle, thus offering a common mechanism, for the liberation of energy from all foodsuffs. Biochemistry has advanced in recent years on several fronts. The primary polypeptide structure of proteins proposed by Emil Fischer was confirmed by the brilliant researches of Sanger who established the complete amino acid sequence of the protein hormone insulin.

du Vigneaud proved the structure of octapeptides of the posterior pituitary by direct synthesis. The studies of Pauling and Corey gave the concept of a secondary structure of protein molecules in the form of a helix. Waston and Crick proposed the double-stranded structures of DNA molecule. The base sequences in messenger that code for each of the amino acids is now available as a result of researches of Nirenberg, Ochoa and Khorana. The base sequence of transfer RNA molecules specific for different amino acids was ascertained by many workers such as *Holley, Medison, Zachan* and others.

A final and accurate list of the base sequences in messenger RNA that code for each of the amino acids was made available because of the brilliant researches by *Nirenberg,* who employed synthetic nucleotides as messenger molecules to identify the coded base sequences for each of the amino acids. A field of biochemistry that lips stimulated lively interest is biochemical genetics. The work done by Jacob and Wollmar, Muller, Beadle and Tatum established that DNA is coded with respect to its abilities (genes) to direct the synthesis of specific protein. Perhaps the most exciting area of present research deals with mechanisms for the regulation of the synthesis of important cellular compounds.

The phenomenon of feedback inhibition of enzyme activity by the end product of a reaction sequence illustrates a self-regulating mechanism. Enzyme induction and repression the acceleration or inhibition of synthesis of an enzyme are described. These discoveries give rise to a hypothesis proposed by *Jacob* and *Monod,* suggesting that the DNA molecules consist of areas in which genes have been maintained in air inactive state (by repressors) until they need to get activated for the production of messenger RNA molecules.

Metabolic Mill

Chemical changes in living organisms usually occur in a series of steps which are referred to as *metabolic pathways.* The compounds formed in this series of reactions are called *intermediates.* Metabolic pathways have been classified as *anabolic, catabolic* and *amphibolic.* In *anabolic* or biosynthetic a relatively simple compound is built up into a more complex molecule. Such pathways therefore build up or synthesize cell constituents. *Catabolic* or degradative pathways, on the other hand, result in the breakdown of compounds or decrease in molecular complexity. Relatively high molecular weight, compounds are degraded to lower molecular weight compounds. Energy released during catabolism is often stored in the form of ATP or some other high energy compounds.

The products of a catabolic pathway are excreted if toxic. Some products may be diverted into anabolic pathways. In some cases it is difficult to decide whether a pathway is anabolic or catabolic. Such pathways have been termed as *amphibolic* (or diabolic). The energy-yielding pathways of glycolysis, pyruvate oxidation and the Krebs cycle are sometimes referred to as catabolic or degradative pathways. However, several biosynthetic pathways branch off from the intermediates of these pathways, The pathways are therefore better designated as amphibolic. The term catabolic would apply to the short sequences that convert amphibolic intermediates to fermentation end products. The pathways from the amphibolic intermediates to the end products would be the purely anabolic pathways. The breakdown products, of the three main classes of foods, carbohydrates, proteins and facts, can participate in glycolysis and the Krebs cycle. These products exist in 'metabolic pools' in the cytoplasm, and can be utilized for cell respiration.

Thus, intereonversion of protein, carbohydrates and fats can occur through the glycolytic pathway and the Krebs cycle. This metabolic interconversion is called the *metabolic mill.*

(1) *Carbohydrates* (polysaccharides) are hydrolysed during the process of digestion to for *monosaccharides.* Of these, the hexose sugar glucose is the key compound in respiration. Other sugars like *galactose* and *fructose* can also serve as sources of energy, and are readily convertible into glucose. Glucose is metabolized through the glycolytic pathway and Krebs cycle to yield energy in the form of ATP.

(2) *Proteins* are digested into their constituent amino acids. These amino acids undergo deamination and enter at various points, in the glycolytic pathway and the Krebs cycle. During deamination, the amino (-NTH_2) group of the amino acid is removed. Thus, *alanine* on deamination forms *pyruvic* acid, and glutamic acid forms *&-ketoglufaric* acid.

The amino, acids alanine, serine, cysteine, cystine, cystine, glycine, threonine, hydroxyproline, valine and methionine enter the glycolytic pathway at the *pyruvic acid* stage. Aspartic acid is converted into *oxaloacetic* acid, and enters the Krebs cycle. Methionine, leucine, isoleucine and valine enter the Krebs cycle at the *succinyl CoA* stage.

Glutamic acid, glutamine, arginine, ornithine, histidine, proline, hydroxyproline and lysine enter the Krebs cycle at the & *ketoglutaric* acid stage. Phenylalanine, tyrosine, tryptophan, lysine, leucine and isoleucine ewer as *acetoacetyl CoA*. The pathway can be reversed under certain conditions. Thus by addition of an amino (-NH_2) group *(amination)* &-ketoglutaric acid forms *glutamic acid*. Such amino acids which are synthesized by the cell are called *non-essential amino* acids, They are formed by the cell itself, and need not be taken in the food.

On the other hand amino acids which are not synthesized by the cell, but must be obtained from food, are called *essential amino acids.*

(3) *Fats* on digestion are broker, down to glycerol and *fatty acids.* Glycerol enters the glycolytic pathway at the *triose phosphate stage.* Fatty acids undergo *B-oxidation,* by which two carbon fragments of acetyl *CoA* are split off at a time from the fatty acid chain. This ultimately results in complete oxidation of the fatty acid.

The cell is therefore not entirely dependent on outside sources for its materials. Under conditions of carbohydrate scarcity, the, breakdown products of proteins and fats car also be utilized for providing energy to the cell. When carbohydrates are in excess, fats can be synthesized from them. Thus eating of excess of potatoes (carbohydrate) results in fatness. Conversely, after prolonged fasting a person becomes thin, as fat reserves are utilized for cell respiration. This permits a degree of flexibility in the biochemical system of the cell.

Bioenergetics

The study of energy changes accompanying biochemical

reactions is termed as bioenergetics or biochemical thermodynamics. These reactions are accompanied by the liberation of energy as the reacting system moves from a higher to a lower energy level. Generally the energy liberated is in the form of heat. But most of biochemical reactions are isothermic.

Therefore, it is not possible to use the heat liberated to drive the vital processes that require energy. These processes *(e.g.,* synthetic reactions, muscular contraction, nerve conduction, and active transport) obtain energy by chemical or coupling to oxidative reactions. In the living systems, the energy is stored in the form of energy rich or high energy compounds which are mainly phosphates containing molecules. Whenever the energy is needed by metabolic process, these high energy compounds which are mainly phosphates containing molecules.

Whenever the energy is needed by metabolic processes, these high energy compounds undergo decomposition and the energy liberated is utilised. Further, the syntheses of these compounds take place as result of various metabolic activities. For example the, oxidation of NADH in the mitochondrion is the main reaction.

Importance of Bioenergetics

Suitable fuel is required to provide the energy that enables the animal to carry out its normal process. How the organism obtains this energy from its food is basic to the understanding of *normal nutrition and metabolism.* Death from *starvation* occurs when available energy reserves are depleted, and certain forms of malnutrition are associated with energy imbalance *(marasmus).*

The rate of energy release, measured by the metabolic rate, is controlled by the *thyroid hormones, whose malfunction is a cause of disease.* Storage of surplus energy results in obesity, one of the most common diseases of occidental society.

Energy and its Various Forms

Energy of a system may be defined as the capacity of a system to do work. This capacity may be bound in the molecules. The bound form of energy is termed as *potential energy.* Potential energy is generally measured' in terms of work that could be undertaken when the energy were released. Potential energy may be of various types such as electrical, chemical, photic (light), atomic or positional.

In the biological world, the main source of potential energy is the solar energy which gets fixed up in the form of chemical energy of carbohydrate molecules during photosynthesis in green plants. Other organisms have to depend directly or indirectly upon this process (photo synthesis) for a constant supply of energy to maintain their structures and to perform their numerous functions. Energy is not only required for mechanical work, maintenance of body temperature and osmotic work but also to drive numerous synthetic reactions. The molecules in motion possess capacity to carry out work.

The energy possessed by a molecule or a body by virtue of its motion is termed as *kinetic energy.* This type of energy depends upon the mass of body and its velocity of motion in accordance to the following relation:

$$K.E = \frac{1}{2} mv^2$$

where K.E. represents the kinetic energy, in the mass of the body and v the velocity of the body.

Other forms of energy are as follows:

— Free energy.

— Internal energy.

The Joule is the unit of energy and is numerically equivalent to 0.239 calorie. One calorie is equal to 4.184 Joules (J) and one Kilocalorie is equal to 4.184 kilojoules (KJ).

Principles of Thermodynamics

Biologic Systems conform to the general laws of thermodynamics. Many quantitative observations carried out on the interconversion of various forms of energy have helped scientists to the formulation of two fundamental laws of thermodynamics, the first and second. These laws have been found to be useful in understanding:

(a) the direction of a reaction, whether from left to right or vice versa,

(b) the accomplishment of work, whether useful or not, and

(c) whether the energy for driving a reaction must be delivered from an external source.

1. *First Law of Thermodynamics:* The principle of Conservation of energy which is based on cumulative experience of ages can be stated as:

   "Energy may be converted from one form to another, but it is impossible to create or destroy it".

Previously, different attempts were made to disprove this principle by constructing a machine that may continue to operate without any expenditure of fuel and may thus create energy out of nothing *(Perpetual motion machine of first kind)*. But all, attempts failed. The net result is that the principle of conservation of energy is universally accepted and is retitled as the first *law of thermodynamics*. There are various ways of enunciating the first law of thermodynamics. Here some of these are given as follows:

(i) *First statement:* It may be put as:

"Energy of an isolated system must remain constant, although it may be transformed from one form to another.

"As heat is given out to the surroundings in exothermic reactions, the internal energy of reacting system decreases while the energy of surroundings increases correspondingly. Therefore, the total energy of the system and the surroundings (i.e., isolated system) taken together remains unchanged. As heat is absorbed from the surroundings in exothermic reactions, the internal energy of the reacting system increases while the energy of the surroundings decreases correspondingly. Therefore, the total energy of the system and the surroundings taken together remains unchanged.

(ii) *Second statement:* It 'is as below:

"Energy in one form, if it disappears, will make its appearance in an exactly equivalent quantity in another form.

"For example, coal contains inherently a lot of internal energy. During burning this energy functions as chemical energy and yields heat. The beat is converted into the mechanical energy of the steam, which moves the engine to move a train. In this example, energy is disappearing in one form and appearing in another form in an exactly equivalent quantity.

Mathematical formulation of the first law: Consider a system whose state on absorbing a quantity of beat 'q' changes from A to B. This heat is used in two ways:

(i) In increasing the internal energy of the system, i.e.

$$\Delta E = E_B - E_A$$

where E_A and E_B denote the internal energies associated with the system in its states A and B respectively, and

(ii) To do some external work *'w'* by the system on its surroundings.

Differential form of the first law. For an infinitesimal process, Eq. (1) may become as follows:

$$dE = \delta q + \delta w \qquad (1)$$

where an infinitesimal process may be defined as a process involving only infinitesimal changes in thermodynamic variable defining the system. The differential is distinguished in notations by *d* from δ. This can be understood by the following discussion. *Change in internal energy is independent of the path taken:*

The change in internal energy is dependent on path by which it is accomplished and the previous history of the system. Thus, change in internal energy is an exact differential. So in mathematical terms, an exact differential is always denoted by a notation *d, e.g., dE* in equation (1).

Nature of δq and δw:

The *'q'* and *'w'* are not state functions because changes in their magnitude for a process depend upon:

(i) the path by which the change is accomplished, and

(ii) the previous history of the system.

The work done in passing from state A to state B by four different paths. From the above discussion, it follows that *w* and *q* are not definite quantities. Mathematically, infinitesimal changes in q and w are not exact differentials but are inexact differentials. In mathematical language, we always write the inexact differential by δq, δw, etc., [Eq. (1)].

Sign of q: When heat is absorbed or gained by the

system, q is + *ve*. When heat is evolved or lost by the system, q is -ve.

Sign of w: When work is done by the system, the value of w is +ve. When work is done on the system, the value of w is-ve

Sign of ΔE: When E_2 (the energy in the final state of the system) is greater than E_1 (the energy in the initial state of the system), the increase in energy (ΔE) is positive but when E_2 is less than E_1, then ΔE is negative and represents decrease in energy.

Significance of the first law of thermodynamics: This law establishes an exact relation between heat and work. It tells us that a certain quantity of heat will produce a definite amount of work and *vice-versa*. The first law of thermodynamics denies that work or energy can be produced from nothing. Work cannot appear without disappearing of heat.

Modification: The discovery of Albert Einstein that mass is convertible into energy, however, makes it essential to make a change in the definition of the first law of thermodynamics. It is possible to convert mass into energy in accordance to the following equation:

$$E = mc^2$$

where E is the energy produced by the destruction of mass m and c is the velocity of light. The modified first law of thermodynamics may be stated as follows:

> "The total. mass and energy of an isolated system remains unchanged."

Final remark.. As matter to energy or energy to matter conversions are of little importance in the biological systems, it means that the first law of thermodynamics has no valid applications in biochemistry.

Limitations of First Law of Thermodynamics: There are three main limitations of first law of thermodynamics, These are as follows :

(a) The first law of thermodynamics simply establishes equivalence between different form of energy. *However, this law does not tell us under what conditions and to what extent it is possible to bring about conversion of one form of energy into the other.* This can he understood from the following examples:

i) We know that heat is flowing from a hot to cold body. First law simply tells us that the quantity of heat lost by the hot body is equal to the heat gained by the cold body. It never tells us that heat can only flow from the hot to the cold body and not in the reverse direction.

ii) Zinc dissolves spontaneously in sulphuric acid, a definite amount of energy being liberated. The reverse process is not spontaneous. Without external influence, the reaction proceeds in only one direction.

From the above examples we can predict that there must be some laws besides the first law which should from the basis of predicting the direction of a certain reaction to occur, and, if so, and to what extent. This, however, comes within the scope of the second law of thermodynamics.

b) The first law of thermodynamics is a qualitative statement which does not prelude the existence of either a heat engine or a refrigerator. The first law does not contradict the existence of a 100% efficient heat engine ora self-acting refrigerator. In practice, these two are not attainable. These phenomena are recognised and this led to the formulation of a law governing these devices. It is called the second law of thermodynamics.

c) The first law fails to explain why chemical reactions do not proceed to completion. For example, consider the reaction $CO(g)+H_2O(g)$ $CO_2(g)+H_2(g)$. If originally one mole of CO and one mole of H_2O were in the reaction vessel, then if the reaction went to completion, one mole of CO_2 and one mole H_2 would be produced. If the reaction vessel is examined at equilibrium, the yield is less than 100%. However, this comes within the scope of second law of thermodynamics.

Internal Energy: It is the energy associated with a system by virtue of its molecular constitution and the motion of its molecules. The contribution of energy *due* to molecular constitution is know it as *internal potential energy* and the contribution of energy due to the motion of molecules is known as *internal kinetic* energy. Hence the total internal energy of a system is given by the sum of two types of energies. i.e.,

$$\frac{\text{Total internal}}{\text{energy}} = \frac{\text{Internal potential}}{\text{energy}} + \frac{\text{Internal kinetic}}{\text{energy}}$$

A fixed quantity of any substance, under a given set of conditions, has got a definite amount of internal energy associated with it. This amount is different for different substance. It is, therefore, regarded as *an extensive property*.

Internal energy is symbolized by E or U.

2. *Second Law of Thermodynamics*: All the natural processes tend to have a tendency to go equilibrium. The second law of thermodynamics is related to the equilibrium state of a process.

Second law of the thermodynamics is generalisation of certain experiences about heat engines and refrigerators, It has been stated in a number of ways. but all the, statements are logically equivalent to one another.

(a) *Kelvin-Planck statement:* "It is impossible to obtain a continuous supply of work from a body or engine that can transfer heat from a single heat reservoir".

The above statement is a negative statement. By this statement, it means that a single reservoir at a single temperature cannot continuously transfer heat into work. It means that there should be two reservoirs for any heat engine. One reservoir (called the *source)* is taken at a higher temperature while the other reservoir *(called the sink) is* taken at a lower temperature. According to this statement, zero degree absolute temperature is not attainable as no heat is rejected to the sink at zero degree Kelvin.

Suppose an engine works between any temperature higher than zero, degree Kelvin. It implies that it uses a single reservoir which controdicts the Kelvin-Planck's statement of the second law. Similarly, it can be said that no engine can be 100% *efficient.*

(b) *Clausius statement*. Heat flows from *hot* to cold body due to difference of temperature. However, in a refrigerator, heat gets absorbed from a body at lower temperature and gets rejected to a body at higher temperature. Here heat is being flown through some outer agency, i.e., it is not feasible to cause heat to flow from a colder to a hotter body without absorbing from some external source. This led Clausius to state the second law of the thermodynamic in the following words : "It *is impossible to transfer for a self acting machine, unaided by any external agency to convert heat continuously, from one body at lower temperature to a body at a higher temperature.*

(c) *Planck statement:* "It is impossible to construct a device that will work in a complete cycle and convert heat into work without making any change in the surroundings.

(d) *Thomson statement:* In a heat engine, the working substance takes heat from hot body; a part of it is converted into work while remaining is given to the sink. No engine has so far been constructed which continuously takes heat from a single body and changes the whole of it to work without making any change in the system. Hence the presence of a cold body is a necessity for converting heat into work. This led Thomson to state that *"it is impossible to obtain a continuous supply of work by cooling a body to a temperature lower than that of coldest of its surroundings.*

(e) *Diffusion* of *a gas always takes place from a region of higher to one at a lower pressure, until the pressure becomes uniform.*

Thus, the second law of thermodynamic may be stated in various ways. All the statements, appear to be different from one another. But a closer look at them, reveals that they are all the modified forms of the same fundamental idea that, *all forms of energy are convertible into heat but the heat so obtained cannot be converted into other forms of energy by any process.* The difference between the first and second laws of thermodynamics lies in the fact that *the first law* is concerned with the accounting of the various kinds of energy involved in a given process, while the *second law* is concerned with the availability of the energy of a given system for doing useful work,

Entropy

The second law of thermodynamics introduces the concept of *entropy* which may be regarded as the amount of energy unavailable for conversion into work. The term entropy was produced by Clausius (1865) and is denoted by symbol S. In simple words, entropy of a system may be regarded as the amount of energy unavailable for conversion into work. It is, however, very difficult to define actual entropy of a system precisely. Therefore, more generally the change in the entropy of a system is defined rather than the absolute entropy of the system.

The increase in entropy dS in the course of an infinitesimal change is equal to δq_{rev}, the heat absorbed when the change is carried out in a reversible manner divided by the absolute temperature T, i.e.

$$dS = \frac{\delta q_{rev}}{T} \quad (2)$$

Although δq_{rev} is not an exact differential, it has a definite value for a reversible, isothermal change, and dS as defined by equation (1) is an exact differential. The entropy S of a system is thus a state function.

If heat change takes place at different temperatures, than we have

$$dS = \frac{\delta Q_1}{T_1} + \frac{\delta Q_2}{T_2} + \frac{\delta Q_3}{T_3} + ... = \Sigma \frac{\delta Q_{rev}}{T} \quad (3)$$

Eq. (1) may be written as follows

$$dS = \frac{\delta Q_{rev}}{T}$$

or

$$S = \frac{\delta Q_{rev}}{T} \quad (4)$$

From equation (3) it *means that "the entropy difference between two states of a given system is equal to the summation of all the* Q_{rev}/T *terms of all the steps which are necessary to bring about a reversible transformation between these two* *states, where q_{rev}, denotes the heat absorbed reversibly and T the corresponding temperature.*

Units of entropy: For a finite change at constant *temperature dS becomes ΔS, δq_{rev}* becomes q_{rev} and Eq. (1) becomes as follows:

$$S = \frac{q_{rev}}{T}$$

$$S = \frac{\text{Heat change}}{\text{Absolute temperature}}$$

$$= \frac{J}{K} = JK^{-1}$$

ΔS is an extensive property and its value, therefore, depends upon the amount of substance involved. Therefore, the unit of entropy will be JK^{-1} mol^{-1} (S.I. units). In both physical and biological world, the entropy is always changing. Whenever there occurs spontaneous change in process the entropy increases.

Physical significance: (i) A characteristic of a spontaneous process (such as the flow of heart in metal base from the hot to the cold end, diffusion of one gas into another, the dissolution of a solid by a solvent etc.) is that it is accompanied by an increase in the 'disorder' or randomness' of the molecules of entropy.Hence entropy may be regard as" *a measure of disorder or randomness of the molecular arrangement in a system*

(ii) A spontaneous process always proceeds from a less probable state to a more probable state by the interplay of natural forces. We know that a spontaneous process is always accompanied by an increase in the entropy, we can therefore identify the entropy as a function of the probability of the state. Since we know that both the entropy, and the thermodynamic probability increase simultaneously in a process we can therefore consider the state of equilibrium as the state of maximum probability. It shows that a system, when left to itself, tends to attain a state of equilibrium,

Free Energy

In biological systems, the free energy is the most important type of energy. It is denoted by *G.* It may be regarded as a measure of the potential energy of the substance but it is a quantity that cannot be measured directly. However, its change accompanying chemical reaction or a process can he measured. It is denoted by ΔG. *The change in free energy (ΔG is that portion of the total energy change in a system which is available for doing work., i.e., it is the useful work.)*

The change in free energy ΔG of a reaction is calculated by taking the difference between the sum of the free energies of the products and that of reactants.

$$A + B \qquad C + D \tag{5}$$

$$\Delta G = \text{Free energy} = (G_c + G_D) - (G_A = G_B) \tag{6}$$

When the free energy change in a reaction is negative (ΔG negative), the reaction is called exergonic and when the free energy change is positive, (ΔG positive), the reaction is called endergonic. An example of exergonic precess is the process of respiration while an example of endergonic process is the process or photosynthesis. It is often said that the reactions with a negative ΔG can occur spontaneously but the sign and magnitude of ΔG give no information on the rate of a reaction. Thus, for the complete oxidation of a glucose according to the equation.

$$C_6H_{12}O_6 + 6O_2 \longrightarrow 6CO_2 + 6H_2O$$

ΔG = 2870 kJ per mole of glucose; in spite of this large negative ΔG, glucose is quite stable in air and the rate spontaneous reaction being negligibly small. Further if ΔG is zero, a state of chemical equilibrium exists and no net chemical change takes place.

***Relation between* ΔG *and* ΔS**

As the entropy charges of chemical reactions could not be readily measurable, the entropy is not regarded as criterion whether a biochemical process can take place spontaneously or not. Furthermore, for spontaneity, both entropy changes (that of the surroundings and of the system) should be known.

These difficulties could be overcome by using a different thermodynamic function, called the *free energy, which is* denoted by the symbol G. T*he symbol* G has been given in honour Josiah Gibbss (1839-1903), an American physical chemist. who single-handedly created a large portion of chemical thermodynamics. In 1878, Gibbss introduced the free energy function by combining the firs', and second laws of thermodynamics in the form of following equation:

$$\Delta G = \Delta H - T\Delta S \quad (7)$$

where, ΔG = the change in free energy of a creative system

ΔH = the change in heat content or enthalpy of this system

T = the absolute temperature at which the process is taking place, and

ΔH = the change in entropy of this system.

In fact, this equation gives quantitative interrelationship between the changes in energy, heat and entropy in chemical reactions taking place at constant temperature (T) and pressure (P), the conditions prevailing in biological systems. The term $T\Delta S$ is that fraction of AH which cannot be converted into useful work.

The ΔG reveals the free energy change or the theoretically available useful work. Naturally, in most cases the system has been 'inefficient' and not all of the theoretically available work can be utilised. However, the properties of the surroundings do not take part into this process:

The enthalpy change, ΔH is given by the following equation:

$$\Delta H = \Delta E + p\Delta V \quad (8)$$

where, ΔE = the change in internal energy of a reaction, and

ΔV = the change in volume of this reaction.

As the volume changes, DV has been found to be small for nearly all, biochemical reactions, hence DH has been nearly equal to the change in internal energy, DE. Therefore, Equation (3) becomes as follows:

$$\Delta G \approx AE - T\Delta S \quad (9)$$

Thus, the change in free energy of a reaction, ΔG will depend both on the change in internal energy and on the change in entropy of the system. The ΔG has been a valuable criterion in determining whether a reaction can take place spontaneously. Thus,

(i) I ΔG if negative, the reaction takes place spontaneously with loss of free energy, i.e., it is exergonic. If, in addition ΔG is of great magnitude, the reaction goes virtually to completion and has been essentially irreversible.

(ii) If, however, ΔG is positive, the reaction proceeds only if free energy can be gained, i.e., it is endergonic. If, in addition, ΔG is of high magnitude, the system is stable having little or no tendency for a reaction to take place.

(iii) If ΔG is zero, the system is at equilibrium and no net change takes place. It means that the system is stable with little or no tendency for a reaction to occur.

With regard to the free energy change DG of a reacting system two more points have to be understood.

1. The ΔG of a reaction is dependent only on the free energy of the products minus that of the reactants. *The ΔG of a reaction is independent of the path of transformation. Obviously,* it means that the mechanism of a reaction has no effect on ΔG. As an instance, the value of ΔG has been the same for the oxidation of glucose to CO_2 and H_2O whether it occurs by combustion or by a series of enzyme-catalysed reactions.

2. The value of ΔG gives no information about the rate of a reaction. A negative ΔG reveals that a reaction can take place spontaneously, but it does not signify that it will take place at perceptible rate. As we already know that, the rate of a reaction rather depends on the free energy of activation ($\Delta G = \neq$, *which* is unrelated to ΔG.

In order to get a feeling for the magnitude of changes in various forms of energy, an actual example of aerobic oxidation of glucose may be taken into consideration. Pie living cells carry out oxidation of glucose in the presence of oxygen to CO_2 and H_2O at constant temperature and pressure.

$$C_6H_{12}O_6+6O_2 \longrightarrow 6CO_2+6H_2O$$

If we assume that the temperature is 25°C (or 298 K) and the pressure is 1.0 mm. (or 760 mm Hg), which are standard conditions in thermodynamic calculations, the following energy changes occur per molecule of glucose oxidised:

$\Delta G = -686{,}000$ cal/mol (i.e., the free energy of the reacting molecules has decreased)

$\Delta H = -673{,}000$ cal/mol (i.e., the reacting molecules have released heat)

Eq. (1) may also be written as:

$$\Delta S = \frac{\Delta H - \Delta G}{T} \qquad (10)$$

On substituting the above values in Eq. (4), we get

$$\Delta S = \frac{-673{,}000 - (-686{,}000)}{298}$$

$$= 44 \text{ cal/deg}$$

$$= 184.096\text{j/deg}$$

Relationship between Standard Free Energy Change and Equilibrium Constant

In a model reaction

$$A + B \qquad C + D \qquad (11)$$

the free energy change, AG of this reaction will be given by

$$\Delta G = \Delta G^\circ + RT \log_e \frac{[C]\,[D]}{[A]\,[B]} \qquad (12)$$

where, ΔG° = Standard free energy change

R = Gas constant

T = Absolute temperature

[A], [B], [C] and [D] = Molar concentrations (i.e., activities) of the reactants.

The standard free energy change (ΔG°) of a chemical reaction may be defined as the free energy change when the reactions and products are in their standard states (298 K, I atm pressure).

Thus, the ΔG of a reaction depends on the nature of the reactants (expressed in ΔG° term) and on their concentrations (expressed in logarithmic terms), as shown in Eq. (2). At equilibrium, $\Delta G = 0$. Equation then becomes as follows:

$$\Delta G = \Delta G^\circ + RT \log_e \frac{[C]\,[D]}{[A]\,[B]} \qquad (13)$$

$$\Delta G^\circ + RT \log_e \frac{[C]\,[D]}{[A]\,[B]} \qquad (14)$$

The equilibrium constant under standard conditions, K'_{eq} for the reaction A + B C + D, will be given as follows:

$$K'_{eq} \qquad \frac{[C]_{eq}\,[D]_{eq}}{[A]\,[B]}$$

At Equilibrium, there is no change in free energy (ΔG= 0); therefore, Eq. (4) becomes as follows:

$$0 = DG^\circ + RT \ln \frac{[C]_{eq}\ [D]_{eq}}{[A]_{eq}\ [B]_{eq}}$$

By changing logarithms to the base 10, the above equation becomes as follows:

$$0 = \Delta G^\circ + 2.303\ RT \log K'_{eq} \tag{15}$$

$$\Delta G^\circ = -2.303\ RT \log K'_{eq} \tag{16}$$

Eq. (16) reveals that if it is possible to determine equilibrium constant for any reaction, the standard free energy change can also be calculated. Conversely if ΔG° is known, the actual free energy ΔG for any reaction can also be calculated.

Eq. (16) can be put in the following form:

or $$K'_{eq} = 10 \tag{17}$$

Substituting $R = 1.98 \times 10^{-3}$ kcal mol^{-1} degree^{-1} and T = 298° K (corresponding to 25°C) gives

$$K'_{eq} = 10$$

when $\Delta°G$ has been expressed in kcal/mol. Hence the standard free energy, ΔG° and the equilibrium constant, K'_{eq} are related by a simple expression. For example, a change in equilibrium constant by a factor of 10 causes a change in standard free energy of 5.69 kj/mol at 25°C (refer Table 1). At 37°C, however, the change in standard free energy would be of 5.69 kj/mol. Values of ΔG° may be expressed in joules or calories per mole.

If ΔG° is negative (i.e., the products have less free energy than the reactants), the reaction will take place to form the products under standard conditions, as all chemical reactions will tend to go in that direction causing a decrease in the free energy of the system. If ΔG° is positive (i.e., the products of the reaction contain more free energy than the reactants), the will proceed in the reverse direction if we start with 1M concentration of all components. To be more explicit, reactions having a negative ΔG° take place forward in the direction written when they start with all reactants and products at 1.0 M, until they reach equilibrium. Reactions with a positive ΔG° will take place in the reverse of the of the direction written when they start with all components at 1.0 M. Table 2 these relationships.

Table 2. Numerical relationship between equilibrium constants of chemical reactions and their standard free energy changes at 25°C

K'_{eq}	*ΔG° (kcal/mol)*	*ΔG° (kj/mol)*
10^{-5} or 0.00001	+6.82	+28.53
10^{-4} or 0.0001	+5.46	+22.84
10^{-3} or 0.001	+4.09	+17.12
10^{-2} or 0.01	+2.73	+11.42
10^{-1} or 0.01	+1.36	+5.69
1	1	0
10^{1} or 10	−1.36	−5.69
10^{2} or 100	−2.73	−11.42
10^{3} or 1,000	−4.09	−17.12
10^{4} or 10,000	−5.46	−22.84
10^{5} or 1,00,000	−6.82	−28.53

Table 3. Relationship among K'_{eq} ΔG° and the directions chemical reactions understand conditions.

When K'_{eq} is	*ΔG is*	*Starting 1M components, the reactions*
>1.0	Negative	Proceeds forward
1.0	Zero	Remainsat equilibrium
<1.0	Positive	Proceeds backward

Two more points should be understood:

1. Biochemical reactions occur near pH 7.0. Hence, PH 7.0 is conventionally designated the standard pH in biochemical energetics. The standard free energy change at pH 7.0 in biochemical energetics has been designated by the symbol, ΔG°.
2. The basic SI (Syste'me International) unit of energy is joule (J). However, in biology and medicine, beat and energy measurements are equally expressed in calories (cal). The interrelationship between calories and joules is:

$$1.000 \text{ cal} = 4.184 \text{ J}.$$

Biochemist's Standard State

It is important to first distinguish between the standard states employed by physical chemists and biochemists. In physical chemistry, the standard state refers to the situation in which all the reactants and products are at unit molar (or molal) concentration. In biochemistry, we follow the same procedure except that we define the hydrogen-ion concentration for the standard state to lie 10^{-7} M because the physiological pH is about 7. Consequently, the change in the standard Gibbs free energy according to these two conventions will be different for reactions involving uptake or liberation of hydrogen ions. We shall therefore replace ΔG° with ΔG° in discussing biochemical process. Consider the reaction

$$A + B \longrightarrow C + xH^+$$

The Gibbs free energy change (ΔG) for the process is given by

$$\frac{[C]}{1M} \quad \frac{[H^+]}{1M}$$

$$\Delta G = \Delta G^\circ + RT\ In\ \frac{[A]}{1M}\ \frac{[B]}{IM}$$

where 1M represents the physical chemists' standard state of solute in solution. Since the biochemists standards state for H^+ ions is $10^{-7}\,M$, the Gibbs free energy change for the same process is now given by

$$\Delta G = \Delta G^\circ + RT\ In\ \frac{\dfrac{[C]}{1M}\ \dfrac{[H^+]}{10^{-7}\,M}}{\dfrac{[A]}{1M}\ \dfrac{[B]}{IM}}$$

Note that regardless of the convention for the standard state, ΔG must remain unchanged. From the last two equations we obtain. This means that for reactions producing H^+ ions, ΔG° is grater than ΔG° by 40.0 kj per mole of H^+ ions released.Hence the reaction is more spontaneous at pH 7 than at pH 0. On the other hand, if H^+ ion appears as the reactant.

$$C + xH^+ \longrightarrow A + B$$

we can show that

$$\Delta G^\circ = \Delta G^\circ\ - 40.0\ KJ$$

and this reaction will be more spontaneous at pH 0 than at pH 7. For reactions not involving H^+ ions, ΔG° is equal to ΔG°.

Standard Free Energy Changes at pH 7.0 or ΔG°

In order to calculate ΔG° we will consider an example of isomerization of dihydroxy-acetone phosphate (DHAP) to glyceraldehyde 3-phophate (G-3-P). It has been one of the reaction of glycolysis.

At equilibrium, the ratio of glyceraldehyde 3-phosphate to dihydroxy- acetone phosphate has been found to be 0.0475 at 250C (298 0 K) and pH 7.0. Hence

$$K'_{eq} = 0.0475$$

The standard free energy change for this reaction is, then, calculated by using Eq. (1) as follows:

$$\Delta G^0 = -2.303 \text{ RT } \log_{10} K'_{eq}$$

$$= -2.303 \times 1.98 \times 10^{-3} \times 298 \times \log_{10} (0.0475)$$

$$= +1.8 \text{ kcal/mol}$$

If the initial concentration of DHAP has been 2×10^{-4} M and the initial concentration of G-3-P is 3×10^{-6} M' then ΔG can be calculated by using Eq. (2) as follows:

$$\Delta G = \Delta G^\circ + \text{RT } \log_e \frac{[C]\,[D]}{[A]\,[B]}$$

$$= 1.8 \text{ kcal/mol} + 2.303 \text{ RT } \log_{10} \frac{3 \times 10^{-6} \text{ M}}{2 \times 10^{-4} \text{ M}}$$

$$= 1.8 \text{ kcal/mol} - 2.5 \text{ kcal/mol}$$

$$= -0.7 \text{ kcal/mol}$$

$$= -2.93 \text{ kJ/mol}$$

The negative value for ΔG reveals that isomerization of dihydroxyacetone phosphate to glyceraldehyde 3-phosphate can take place spontaneously, when these compounds are present at concentrations mentioned above. Although ΔG has been negative for this reaction, yet the value of ΔG° has been positive. It is important to note that the magnitude of ΔG for reaction (whether smaller, larger or the same as ΔG°) has been found to depend on the concentrations of the reactants. The criterion of spontaneity for a reaction is ΔG and not ΔG°. Table 3 lists the standard free energy changes for some chemical reactions.

Table 4. Standard free energy changes for representative chemical reactions under standard conditions (that is, at 25°C and pH. 7.0)

Reaction type Reactions	ΔG° (*in kcal/mol*)	ΔG° (*in kj/mol*)
Elimination of Water		
Malate → Fumarate+ H_2O	+ 0.75	+3.138
Rearrangement		
Glucose 1- phosphate → Glucose 6- phosphate	–1.74	–7.280
Fructose 6-phosphate Glucose 6- phosphate	–0.40	–1.6736
Hydrolysis		
Esters		
Enthyl acetate +H_2O → Ethanol+ Acetate	–4.7	–19.665
Glucose 6- phosphate+ H_2O Glucose→+Phosphate	–3.3	–13.807
Amides and peptides		
Glutamine + H_2O → Glutamate + NH_4^+	–3.4	–14.225
Glycyglycine +H_2O → 2 Glycine	–2.2	–9.205
Glycosides		
Maltose +H_2O → 2 Glucose	–3.7	–15.481
Lactose H_2O → Glucose+ Galactose	–3.8	–15.899
Acid anhydrides		
Acetic anhydride +H_2O → 2 Acetate	–21.8	–91.211
ATP+ H_2O → ADP+ Phosphate	–7.3	–30.543
Oxidations with molecular oxygen		
Glucose + $6O_2$ → $6CO_2$+ $6H_2O$	–686	–2870.224
Palmitic acid + $23O_2$ → $16CO_2$ +16 H_2O	–2338	–9782.192

When we go through the above table, we find that hydrolysis of esters, amides, peptides and glycosides, as well as rearrangements, and elimination takes place with relatively small standard free energy change. Hydrolysis of acid anhydrides takes place with relatively large decreases in standard free energy, whereas oxidation of organic compounds to $C0_2$ and H_2O occurs with rather huge decreases in standard free energy.

Difference between ΔG and ΔG°

It is better if we understand the difference between the free energy change, ΔG and the standard free energy change, $ΔG^0$ of a chemical reaction. The actual free energy change, ΔG of a chemical reaction has been a function of the conditions of concentration, pH and temperature under which the reaction is occurring. Moreover the, ΔG of an ongoing chemical process always becomes negative, becomes smaller (i.e., less negative) as the reaction proceeds and becomes zero at the point of equilibrium, revealing that no more work can be done by the reaction.

The value of ΔG decreases with time as the reaction proceeds because the actual concentrations of the reactants will be getting smaller and those of the resultants getting larger. Thus, it can be concluded that in every spontaneous chemical or physical process, the free energy of the reacting system always decreases, i.e., ΔG is negative.

On the other hand., the value of standard free energy change. ΔG°. for a chemical reaction has been characteristic and unchanging, and may be positive, negative or zero, depending on the equilibrium constant of the reaction, ΔG° is, thus an immutable constant and predicts in which direction and how far a given reaction will go in order to reach equilibrium when it takes place under standard condition, i.e., when the initial concentration of all components is 1.0 M, the pH is 7.0 and the temperature is 25°C.

Standard Free Energy Values of Chemical Reactions are Additive

An important fact is that the overall free energy change

for a series of reactions has, been equal to the sum of the free energy changes of the individual steps. Let us consider the two consecutive reactions where each reaction is having its own equilibrium constant and a characteristic standard free energy change ΔG°_1 and ΔG°_2

$$A \longrightarrow B \qquad \Delta G_1^\circ = -8 \text{ kcal/mol}$$

$$B \longrightarrow C \qquad \Delta G_2^\circ = +5 \text{ kcal/mol}$$

As the reactions are sequential, the intermediate product B cancels out and the with its standard free energy change, AG°s may be put as follows:

$$\begin{aligned} A \longrightarrow C \qquad \Delta G^\circ_s &= \Delta G^\circ_1 + G_2^\circ \\ &= -8 + (+5) \\ &= -3 \text{ kcal/mol} \\ &= -12.552 \text{ kj/mol} \end{aligned}$$

Under standard conditions, A can get spontaneously converted into B as G is negative. However, the conversion of B into C, under standard conditions has been not thermodynamically feasible. But as the free energy changes are additive, the conversion of A into C has a ΔG° value of −12.552 kj/mol, which obviously implies that A can be converted into C spontaneously under standard conditions. Thus, the above two sequential reactions get coupled by the intermediate product. B. In other words, a thermodynamically unfavourable reaction can be driven by a thermodynamically favourable reaction.

Two sequential steps from glycogen break down in muscles will demonstrate this fact more clearly.

$$\text{Glucose 1- phosphate} \xrightarrow[\text{Glucose phospate}]{\text{Phospho-glucomutase}} \text{Glucose 6-phosphate} \quad \Delta G^\circ_1 = -1.74 \text{ kcal/mol}$$

$$\text{Glucose 6- phosphate} \xrightarrow{\text{isomerase}} \text{Fructose 6-phosphate}$$

$$\Delta G^{\circ}_{1} = +0.40 \text{ kcal/mol}$$

On adding the two reactions, we obtain

$$\text{Glucose 1- phosphate} \longrightarrow \text{Fructose 6-phosphate}$$

This has a standard free energy change value,

$$\begin{aligned} \Delta G^{\circ}_{s} &= \Delta G^{\circ}_{1} + \Delta G_{2}^{\circ} \\ &= -1.74 + (+\ 0.40) \\ &= -1.36 \text{ kcal/mol} \\ &= -5.6902 \text{ kj/mol} \end{aligned}$$

As $\Delta G^{\circ}s'$ is negative, glucose 1 phosphate gets converted into fructose 6-phosphate in the muscles.

Coupled Reactions

In biochemical reactions, two or more reactions may be taking place together in such a way that an exergonic reaction may be used to drive an endergonic one, Such reactions are known as coupled reactions.

The simplest type of coupling may be represented by the equation:

$$A + C \longrightarrow B + D + \text{Heat}$$

The conversion of metabolite A to metabolite B takes place with the release of energy. It is coupled to another reaction, wherein energy is required to convert metabolite C to metabolite D. Many instances of this sort of coupling are seen in metabolism when the energy derived from an exergonic reaction (usually an oxidation) is used to drive an endergonic synthetic reaction. This sort of coupling requires both reactants to contain a common obligatory intermediate.

$$A + C \longrightarrow I \longrightarrow B + D$$

This type of coupling should have a built-in mechanism for biological control of the rate at which oxidative, processes are allowed to take place because the existence of a common obligatory intermediate for both the exergonic and endergonic reactions allows the rate of utilisation of the product of the synthetic path (D) 'to determine by mass action the rate of which A is oxidised. This type of coupling is provided by dehydrogenation reactions which are coupled to hydrogenation by an intermediate carrier.

An alternative method of coupling an exergonic to an endergonic process may involve the synthesis of a new compound of high-energy potential in the exergonic reaction and then this new compound may he incorporated into endergonic reaction thus effecting a transference of free energy from the exergonic to the endergonic pathway. *E* is a compound of high potential energy and E is the corresponding compound of low potential energy. The main advantage of this mechanism of coupling is that E unlike 1 in the previous mechanism may not be structurally related to A, B, C or D.

This would make E to serve as a transducer of energy from a wide range of exergonic reactions to an equally wide range of endergonic reactions or process. In the living cell, the principal high-energy intermediate or carrier compound (designated ~E) is *adenosine triphosphate or ATP*.

Energy Rich Compounds and Energy Coupling

In order to maintain living processes, all organisms must get free energy from their environment. The autotrophic, organisms get supply of free energy by coupling their metabolism to some simple exergonic process in their surroundings. For example, green plants utilise the energy of sunlight, On the other hand, heterotrophic organisms get free energy by coupling their metabolism to the breakdown of complex organic molecules in their environment. The special carrier of free energy has been adenosine triphosphate (ATP), ATP plays a central role in the transference of free energy from the exergonic (energy yielding) to the endergonic (= energy requiring) processes in the cells.

During breakdown of energy rich foodstuffs or fuel molecules, some of the free energy gets harnessed to make ATP from adenosine diphosphate (ADP) and inorganic phosphate (P_1) a process that requires input of free energy. ATP then donates much of its chemical energy to energy needing processes (biosynetheses transport etc.) by undergoing a breakdown to ADP and P_1. A large number of coupled reactions involve ATP or another nucleoside triphosphate groups.

The importance of phosphates in the intermediary metabolisms was realised by Fritz in 1941 and he introduced the concept of "high-energy phosphates" and "high-energy phosphate bond".

High-Energy Phosphates

ATP is frequently described as an *energy-rich* compound with bounds called high -energy bounds. *A high energy bond is one that liberates a large amount of energy when hydrolyzed (>5 kcal/mole).* Two such high energy groups exist in the ATP anhydride linkage. Let's use a block diagram of ATP so that we can focus our attention on its high-energy groups. The anyhydrides can be hydrolzed. Upon hydrolysis of 1 mole of ATP to adenosine diphosphate (ADP), about 7300 cal (30, 500 J) are liberated under the operating conditions of the cell. P_1 indicates inorganic phosphate, all forms of phosphate not combined with a carbon compound.

ATP can also hydrolyze exothermically to produce adenosine monophosphate (AMP) and phyrophosphate (PP_1).

Thus, by exothermically hydrolyzing ATP to ADP or AMP, a cell liberates energy which can be used to power endothermic processes. The hydrolysis of ADP to AMP is exothermic but does not occur frequently in cellular metabolism. In order to indicate the presence of the high energy phosphate group, Lipman employed the symbol NP to indicate *high energy phosphate bond.* This symbol is indicative of the fact that the group attached to the bond, when transferred to an appropriate acceptor, results in transfer of the larger of quantity energy. Due to this reasons tire group transfer potential is used to indicate "high-energy bond".

For Instance, ATP has two high energy phosphate groups and ADP has one while the phosphate bond in AMP (adenosine monophosphate) is of the low-energy type because it is a normal ester link. Cells also contain other high-energy compounds in addition to ATP. When hydrolyzed, these compounds also liberate more than 5 kcal/mole.Structural formulas for several of these compounds are given below. Phosphocreating (creatine phosphate) is an important storage compound.

The phosphate group is added directly and reversibly to creative from ATP under the influence of the enzyme creative kinase. These reversible reactions are particularly important in skeletal muscle tissue. During muscular work, ATP is needed for muscle construction. When the muscle is at rest, ATP concentration is high and the forward reaction above is favoured, a phosphate group is transferred to certain , forming phosphocreatine. In this way, creatine phosphate.

During muscular work, ATP concentration decreases as ATP is converted to ADP. The increased ADP concentration causes the reverse reaction above to be favoured: direct phosphorylation of ADP by phosphocreatine occurs.

Free Energy of Hydrolysis of ATP and other Organophosphates: The standard free energy of hydrolysis of a number of biochemically important phosphates has been given in Table 5.

Table 5. Standard free energy of hydrolysis of some organophosphates of biochemical importance

	Compound	ΔG°	
		kJmol^{-1}	kcal/mole
(i)	Phosphoenol pyruvate	-61.1	–14.8
(ii)	Carbamoy 1 Phosphate	–51.4	–12.3
(iii)	1,3-Biphosphogly cerate (to 3 phosphogly cerate)	–49.3	–11.8
(iv)	Creatine phosphate	–13.1	–10.3
	ATP ⟶ ADP + Pi	–36.8	–8.8
(v)	Glucose 1-phosphate	–20.9	–5.0
(vi)	Fluctose 6-phosphate	–15.9	–3.8
(vii)	Glucose 6-phosphate	–13.8	–3.2
(viii)	Glycerol 3- phospate	–9.2	–2.2

From Table 5 it can be seen that the value for the hydrolysis of the terminal phosphate of the ATP of 36.8 kJ (8.8) (kcal) per mol (as it is also for the terminal phosphate of ATP) divides the list into 2 groups.

(i) *Low energy compounds:* This group has ΔG°′ values which are smaller than that of ATP. This group is exemplified by the ester phosphates found in the intermediates of glycolysis.

(ii) *High energy compounds:* The group has ΔG°′ values

which are higher than that of ATP. This group is exemplified by anhydrides (e.g., ATP, ADT, the 1-phosphate of 1, 3-biphosphogy-cerate), enolphosphates (e.g., phosphoenolpyruvate) and phosphogunidines (e.g., creative phosphate, arginine phosphate). Other biologically important compounds that are classified as "high-energy compounds" are thiol esters involving coenzyme. A (e.g., acetyl CoA), acyl carrier ' protein, amino acid esters involved in protein synthesis, Sadenosy Imethionine (active methionine) and UDPG (uridine diphosphate glucose).

Role of High Energy Phosphates as the Energy Currency of the Cells: Due to its middle position in the Table 5, ATP acts as a donor of high-energy phosphate to those compounds which are below it in the Table 4. Similarly, ADP can accept high-energy phosphate to form ATP from those compounds which are above ATP in the Table 5. In reality, an ATP/ADP cycle connects these processes which generate ~ P to those processes which utilise D9. In Table 4, there is another group of compounds which act as storage forms of high-energy phosphate. These are creatine phosphate (phosphagen) and arginine phosphate.

Bioenergetics of Coupled Reactions: In biological systems. ATP serves as a source of energy for many of the endergonic reactions. Also, ATP get synthesised in many exergonic reactions. When an energonic reaction gets coupled with an exergonic reaction through ATP, the overall effect of this coupling involves the transfer of kinetic energy from one system to another. Several enzymes are known which facilitate these coupling reactions.

The, coupling of reactions so as to bring about efficient transfer of energy is quite common in biological systems. This is illustrated by considering a reaction which is the first in the glycolysis pathway. For example, the formation of glucose 6 phosphate from glucose arid phosphoric acid requires energy with $\Delta G° + 13.8$ kJ mol^{-1}

(i) Glucose + Pi $\rightarrow$ Glucose 6-P + H_2O; $\Delta G°' = + 13.8$ kJ mol^{-1}

The above reaction would not proceed as such under physiologic condition. In order to take place, the reaction must be coupled with another reaction which should be more exergonic than the phosphorylation of glucose which is endergonic, Such a reaction is nothing but the hydrolysis of the terminal phosphate of ATP.

(ii) ATP $\longrightarrow$ ADP + Pi ; $\Delta G°' = 36.8$ kJ mol^{-1}

If (i) and (ii) get coupled in a reaction catalysed by hexokinase, there occurs ready phosphorylation of glucose in a highly exergonic reaction that under physiologic conditions is far from equilibrium and thus irreversible for practical purposes.

$$\text{Glucose} + \text{ATP} \xrightarrow{\text{Hexokinase}} \text{Glucose 6.P} + \text{ADP}$$

$$\Delta G° = + 23.2 \text{ kJ mol}^{-1}$$

Many "activation reactions" have been, found to follow this pattern.

This series of coupled reactions may be put as follows:

By using, energy coupling a complex compound could be apparently which violates the second law of thermodynamics. But the algebraic sum of energy of all the components of coupled reaction is negative, i.e., he sum of energy liberated has been found Lo be more than the sum of energy consumed. In biological systems, the hydrolysis of energy rich compounds is generally accompanied by an endergonic reaction O that the energy liberated in the former reaction is used fully without getting wasted. This type of control needs numerous enzyme.; which are present in the cell. The high energy compounds get dissociated in the presence of enzymes.

Adenylate Kinases Allows Interconversion of Adenine Nucleotides

The enzyme adenylate kinase (myokinase) is present in most cells. It catalyzes the Interconversion of ATP and AMP on the one hand and ADP on the other:

$$\text{ATP} + \text{AMP} \xleftrightarrow{\text{Adenylate Kinase}} 2\text{ADP}$$

This reaction has 3 functions:

(1) It allows high-energy phosphate in ADP to be used in the synthesis of ATP.

(2) It allows AMP, formed as a consequence of several activating reactions involving ATP, to be recovered by rephosphorylation to ADP.

(3) It allows AMP to increase in concentration when ATP becomes depleted and act as a metabolic (allosteric) signal to increase the rate of catabolic reactions, which in turn leads to the generation of more ATP.

When ATP Reacts to Form AMP, Inorganic Pyrophosphate (PPi) is Formed.

This occurs, for example, in the activation of long-chain fatty acids.

This reaction is accompanied by loss of free energy as beat, which ensures that the activation reaction will go to the right; this is further aided by the hydrolytic splitting of PP_1, 1 catalysed by *inorganic pyrophosphatase,* a reaction that itself has a large ΔG° OF –27.6 kJ/mol. Note that activations via the pyrophosphate pathway result in the loss of 2 P rather than one P as occurs when ADP and P_1 are formed.

A combination of the above reaction, makes it possible for phosphate to be recycled and the adenine nucleotides to interchange.

ATP Synthesis

Energy is released upon ATP hydrolysis. This means that the reverse process is endothermic and energy is stored in ATP during its formation.

$$7300 \text{ cal} + \text{ADP } P_1 \qquad \text{ATP} + \text{HOH}$$

$$8000 \text{ cal} + \text{AMP} + PP_1 \qquad \text{ATP} + \text{HOH}$$

A logical question is: Where does the energy come from that's used in ATP formation? It arises during catabolism in which cells chemically break down large metabolite molecule into simpler substances of lower energy. A net release of energy occurs and some of the energy is stored by conversion of AMP and ADP to ATP. The formation of ATP by capturing energy liberated during catabolism is called *oxidative phosphorylation.*

The majority of ATP is produced by this process. Oxidative phosphorylation occurs through the action of the respiratory chain, also known as the chain or the electron transport system (ETS). The components of the ETS are found on the inner membrane surface of mitochondria. The enzymes of the ETS function as a sequence in which hydrogens are passed along carriers. Ultimately the hydrogens combine with oxygen to produce water. Hydrogen atoms can be imagined to consist of component hydrogen ions and electrons.

$$H + H = 2e + 2H^+$$

Recall that the loss of hydrogen with its associated

electron from a substance is called oxidation; gain of hydrogen and, its electron is called reduction. The formation of water involves the reaction

$$2H^+ + 2e^- + \frac{1}{2} O_2 \longrightarrow H_2O$$

Therefore, a biological requirement exists for hydrogen ion and electron carriers. Two kinds of carriers are present in the electron transport system: (1) coezymes, which are electron and hydrogen ion carriers, and (2) cytochromes', which are exclusively electron carriers. The coenzymes presently known to participate in the ETS are (1) nicotinamide adenine dinucleotides. (NAD and NADP), (2) flavin nucleotides (FAD and FMN), and (3) coenzyme Q (CoQ).

High Energy Phosphates Play a Central Role in Energy Capture and Transfer

In order to maintain living processes, all organisms must obtain supplies of free energy from their environment. *Autotrophic* organisms couple their metabolism to some simple exergonic process in their surroundings, e.g., green plants utilise the energy of sunlight, and some autotrophic bacteria utilise the reaction Fe^{2+} ¾→ Fe^{3+}. On the other hand. heterotrophic organisms obtain free energy by coupling their metabolism to the breakdown of complex organic molecules in their environment. In all of these processes, ATP plays a central role in the transference of free energy from the exergonic to the endergonic processes.

The importance of phosphates in intermediary metabolism became evident with the discovery of the chemical details of glycolysis and of the role of ATP, adenosine diphosphate (ADP and inorganic phosphate (P_1,) in this process ATP was considered a means of transferring phosphate radicals in the process of phosphorylation. The role of ATP in biochemical energetics was.indicated in experiments demonstrating that ATP and creatine phosphate were broken down during muscular contraction and that their resynethesis depended, on supplying energy from oxidative processes in the muscle. It was not until Lipmann introduced the concept of "high-energy phosphates" and the "high-energy phosphate bond that the role of these compounds in bioenergetics was clearly appreciated. So far we have presented a rather general and qualitative description of how some of the concepts can be applied to help us understand the nature of biochemical reactions.

In this section we consider a more fundamental question: Under what conditions is thermodynamics applicable to biology? The answer to this question is not a trivial one, and at present there is much debate on this subject. To begin with, we note that biochemical reactions are usually discussed in terms of standard free energies (ΔG°). Strictly, the direction of a reaction such as

$$A + B \qquad\qquad C + D$$

is given by ΔG, where

$$\Delta G \; \Delta G^{O} = + RT \ln \frac{[C]\,[D]}{[A]\,[B]}$$

Only in the case where the reactants and products are in their standard states can we employ ΔG^{O}. Thus the fact that the hydrolysis of ATP at 298 K and pH 7 results in a decrease of standard free energy equal to 30.5 kJ mol^{-1} does not necessarily mean that this is also the value for ΔG. In living cells the physiological temperature, pH, concentration of reactants and products, as well as metal ions can vary from system to system, and these factors must affect the value of ΔG.

It is difficult and in some cases impossible to accurately measure the concentration of various, species in solution. Nevertheless, reasonable estimates can be made of the concentrations in many cases and hence of the sign of ΔG.

Another point that has often beep raised is that thermodynamics deals only with, closed systems at equilibrium. Living systems are open systems and they are maintained at steady state rather than at equilibrium. In fact, any cell at equilibrium is dead cell.

The treatment of open systems requires *irreversible thermodynamics,* which is beyond the scope of this book. The interested reader is referred to some important papers on the application of thermodynamics to biology listed. Finally, reactions in a living cell, like any other reactions, can be classified into two categories- those that are thermodynamically controlled and those that are kinetically controlled.

As an example of the former, let us consider the synthesis of a dipeptide from two amino acids:

$$\text{alanine} + \text{glycine} \longrightarrow \text{alanylgycine} \quad \Delta G^{O} = 17.2 \text{ kJ}$$

The corresponding equilibrium constant for this reaction is about 1×10^{3} at 298 K. Clearly, such a process will not proceed by an appreciable extent if left on its own. However, if this process could be coupled to the hydrolysis of ATP by the action of an enzyme, the reaction would proceed from left to right.

Such a reaction is said to be thermodynamically controlled because the reaction itself is not spontaneous, and the energy(or free energy) of reaction must be supplied from an outside source.

A kinetically controlled reaction is one in which the overall ΔG is negative (and hence thermodynamically favourable), but the rate is negligibly small in the absence of appropriate enzyme catalysis.

The phosphorylation of glucose to glucose 6-phosphate (coupled to ATP hydrolysis) is cerainly an exergonic process, but the reaction occurs at a very slow rate in the absence of the enzyme hexokinase.

Chapter 2

Water and Biochemical Reactions

Water has two hydrogens to enter into hydrogen bonds and two unshared pairs of electrons on the oxygen to which other water .molecules can be hydrogen bonded. Each water molecule is involved in four hydrogen bonds -as donor in two and as acceptor in two. Hydrogen fluoride has only one hydrogen to enter into a hydrogen bond as a donor, but it has three unshared pairs of electrons on the fluorine that could bond to other hydrogens.

Hydrogen bonds are important everywhere in biomolecular structures; they link together parts of protein chains of enzymes and the two complementary chains of the DNA double helix. Another unique property of water is its role in the control of acidity within the cell by buffers. A cell's survival depends on strict control of its internal pH. The geometry of the water molecule and its properties as a solvent play major roles in determining the properties of living systems. When electrons are shared between atoms in a chemical bond, they need not be shared equally. Bonds with unequal sharing of electrons are referred to as polar.

The tendency of an atom to attract electrons to

itself in a chemical bond (i.e., to become negative) is called electronegativity. Atoms of the same element, of course, share electrons equally in a bond -that is, have equal electronegativity-but different elements do not necessarily have the same electronegativity. Oxygen and nitrogen are all highly electronegative more so than carbon and hydrogen. In the O–H bonds in water, oxygen is more electronegative than hydrogen, so there is a higher probability that the bonding electrons are close to the oxygen. The difference in electronegativity between oxygen and hydrogen gives rise to partial positive and negative charge, usually pictured as δ and δ^-, respectively.

The O-H bond is thus a polar bond. In situations where the electronegativity difference is quite small, such as in the C-H bond in methane (CH_4), the sharing of electrons in the bond is very nearly equal, and the bond is essentially nonpolar.

Table 1. Electronegativities of selected elements

Element	*Electronegativity*
Oxygen	3.5
Nitrogen	3.0
Sulfur	2.6
Carbon	2.5
Phosphorus	2.2
Hydrogen	2.1

Carbon dioxide is an example. The two C-O bonds are polar, but because the CO_2 molecule is linear, the attraction of the oxygen for the electrons in one bond is cancelled out by the equal and opposite attraction for the electrons by the oxygen on the other side of the molecule.

$$\delta^- \quad 2\delta^+ \quad \delta^-$$

$$O = C = O$$

Water is a bent molecule with a bond angle of 105 degrees, and the uneven sharing of electrons in the two bonds is not cancelled out as in CO_2. The result is that the bonding electrons are more likely to be found at the oxygen end of the molecule than at the hydrogen end. Molecules with positive and negative ends are called dipoles.

Properties of water

The polar nature of water largely determines its solvent properties. Ionic compounds with full charges, such as potassium chloride (KCl, K^+, and Cl^- solution), and polar compounds with partial charges, such as ethyl alcohol (C_2H_5OH) or acetone ($(CH_3)_2C = O$), tend to dissolve in water. The underlying- physical principle is electrostatic attraction between unlike charges. The negative end of a water dipole attracts a positive ion or the positive end of another dipole. The positive end of a water molecule attracts a negative ion or the negative end of another dipole.

The aggregate of unlike charges, held in proximity to one another because of electrostatic attraction, has a lower energy than would be possible if this interaction could not take place. The lowering of energy makes the system more stable and more likely to exist. These ion-dipole and dipole-dipole interactions are similar to the interactions between water molecules themselves in terms of the quantities of energy involved. Examples of polar compounds that dissolve easily in water are small organic molecules containing one or more

(a)

(b)

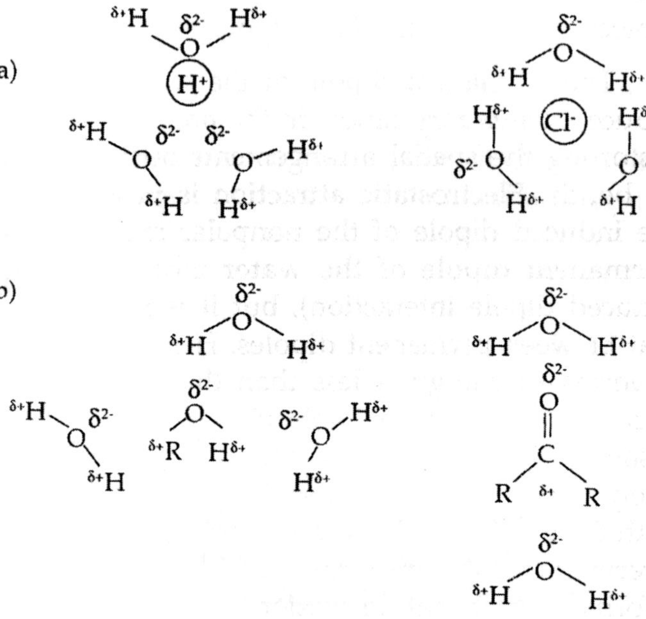

Fig. 1. Ion-dipole and dipole-dipole interactions help ionic and polar compounds dissolve in water.
(a) Ion-dipole interactions with water
(b) Dipole-dipole interactions of polar compounds with water.

electronegative atoms (such as oxygen or nitrogen), including alcohols, amines, and carboxylic acids. The attraction between the dipoles of these molecules and the water dipole makes them tend to dissolve. Ionic and polar substances are referred to as hydrophilic ("water-loving," from the Greek) because of this tendency. Hydrocarbons (compounds that contain only carbon and hydrogen) are nonpolar. The favourable ion-dipole and dipole-dipole interactions responsible for the solubility of ionic and polar compounds do not occur for nonpolar compounds, and so these compounds tend not to dissolve in water. The interactions between nonpolar molecules and water molecules are weaker than dipolar interactions.

The permanent dipole of the water molecule can induce a temporary dipole in the nonpolar molecule by distorting the spatial arrangements of the electrons in its bonds. Electrostatic attraction is possible between the induced dipole of the nonpolar molecule and the permanent dipole of the water molecule (a dipole-induced dipole interaction), but it is not as strong as that between permanent dipoles. Hence, its consequent lowering of energy is less than that produced by the attraction of the water molecules for one another. The association of nonpolar molecules with water is far less likely to occur than the association of water molecules with themselves. However, the points made here about intermolecular interactions will be useful background information for that discussion.

For the moment it is enough to know that it is less favourable thermodynamically for water molecules to be associated with nonpolar molecules than with other water molecules. As a result, nonpolar molecules do not dissolve in water and are referred to as hydrophobic ("water-hating"). (Hydrocarbons in particular tend to sequester themselves from an aqueous environment.) A nonpolar solid leaves undissolved material in water. A nonpolar liquid forms a two layer system with water; an example is an oil slick. A long-chain fatty acid having a polar carboxylic acid group and a long nonpolar hydrocarbon portion is a prime example of an amphipathic substance.

The carboxylic acid group, the "head" group, contains two oxygen atoms in addition to carbon and hydrogen; it is very polar and can form a carboxylate anion. The rest of the molecule, the "tail," contains only carbon and hydrogen and is thus nonpolar. A compound such as this in the presence of water tends to form structures called micelles, in which the polar head groups are in contact with the aqueous environment and the nonpolar tails are sequestered from the water. Interactions between nonpolar molecules themselves are very weak and depend on the attraction between short-lived temporary dipoles and the dipoles they induce. In a large sample of nonpolar molecules there will always be some molecules with these temporary dipoles, which are caused by a momentary clumping of bonding electrons at one end of the molecule. A temporary dipole can induce another dipole in a neighbouring molecule in the same way that a permanent dipole does. The interaction energy is low because the association is so short-lived. It is called a van der Waals interaction. The arrangement of molecules in cells strongly depends on the molecules' polarity, as we saw with micelles.

Importance of hydrogen bond

Hydrogen bonding is of electrostatic origin and can be considered a special case of dipole-dipole interaction. When hydrogen is covalently bonded to an electronegative atom such as oxygen or nitrogen, it has a partial positive charge due to the polar bond, a situation that does not occur when hydrogen is covalently bonded to carbon. This partial positive charge on hydrogen can interact with an unshared (nonbonding) pair of electrons (a source of negative charge) on another electronegative atom. All three atoms lie in a straight line, forming a hydrogen bond.

This arrangement allows for the greatest possible partial positive charge on the hydrogen and consequently for the strongest possible interaction with the unshared pair of electrons on the second electronegative atom.

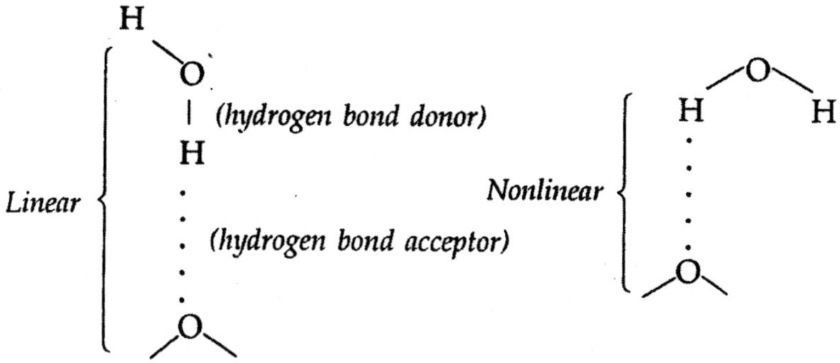

Fig. 2. A comparison of linear and nonlinear hydrogen bonds.

The electronegative atom to which the hydrogen is covalently bonded is called the hydrogen-bond donor, and the electronegative atom that contributes the unshared pair of electrons to the interaction is the hydrogen-bond acceptor (The hydrogen is not covalently bonded to the acceptor.) A consideration of the hydrogen bonding situations in HF, H_2O, and NH_3 can yield some useful insights. Ammonia has three hydrogens to donate to a hydrogen bond but only one unshared pair of electrons, on the nitrogen.

The geometric arrangement of hydrogen-bonded water molecules has important implications for the properties of water as a solvent. The bond angle in water is 105 degrees, and the angle between the unshared pairs of electrons is similar. The result is a tetrahedral arrangement of water molecules. Liquid water consists of hydrogen-bonded arrays that resemble ice crystals; each of these arrays can contain up to 100 water molecules. The hydrogen bonding between water molecules can be seen more clearly in the regular lattice structure of the ice crystal. There are several differences, however, between hydrogen-bonded arrays of this type in liquid water and the structure of ice crystals.

In liquid water, hydrogen bonds are constantly breaking and new ones forming, with some molecules breaking off and others joining the cluster. A cluster can break up and re-form in 10^{-10} to 10^{-11} seconds in water at 25°C. An ice crystal, in contrast, has a more or less stable arrangement of hydrogen bonds, and of course its number of molecules is many orders of magnitude greater than 100. Hydrogen bonds are much weaker than normal covalent bonds. Whereas the energy required to break the O-H covalent bond is 460 kJ mol^{-1} (110, kcal mol^{-1}), the energy of hydrogen bonds in water is about 20 kJ mol^{-1} (5 kcal mol^{-1}). Even this comparatively small amount of energy is enough to drastically affect the properties of water, especially its melting point, its boiling point, and its density relative to the density of ice.

Both the melting point and the boiling point of water are significantly higher than would be predicted for a molecule of this size (Table 4). Other substances of about the same molecular weight, such as methane and ammonia have much lower melting and boiling points. The, forces of attraction between in the molecules of these substances are weaker than the attraction between water molecules, which is due to their hydrogen bonds. The energy of this attraction has to be overcome to melt ice or boil water. Ice has a lower density than liquid water because the fully hydrogenbonded array in an ice crystal is very open, with a lot of empty space between the molecules. Liquid water is less extensively hydrogen bonded, and thus it is denser than ice. Most substances contract when they freeze, but the opposite is true of water. Thus, ice cubes and icebergs float, and expansion resulting from the replacement of liquid water by ice crystals can damage surrounding materials. In cold weather, the cooling systems of cars require antifreeze to prevent freezing and expansion of the water, which could crack the engine block.

Table 2.

Comparison of properties of water, ammonia, and methane

Substance	*Molecular* Weight	*Melting point* (°C)	*Boiling point* (°C)
Water (H_2O	18.02	0.0	100.0
Ammonia (NH_3)	17.03	-77.7	-33.4
Methane (CH_4)	16.04	-182.5	-161.5

The same principle is used in a method of disrupting cells with several cycles of freezing and thawing. Finally, aquatic organisms can survive in cold climates because of the density difference between ice and liquid water because lakes and rivers freeze from top to bottom rather than vice versa. Hydrogen bonding also plays a role in the behaviour of water as a solvent. If a polar solute can serve as a donor or an acceptor of hydrogen bonds, it can form hydrogen bonds with water in addition to being involved in nonspecific dipole-dipole interactions.

Alcohols, amines, carboxylic acids, and esters, as well as aldehvdes and ketones, can all form hydrogen bonds with water, so they are soluble in water. It is difficult to overstate the importance of water to the existence of life on Earth and difficult to imagine life based on another solvent.

Hydrogen bonds other than to water

Hydrogen bonds have a vital involvement in stabilizing the three-dimensional structures of biologically important molecules, including DNA, RNA, and proteins. The hydrogen bonds between complementary bases are one of the most striking characteristics of the double-helical structure of DNA. Transfer RNA also has complex three-dimensional structure characterized by hydrogen-bonded regions.

Hydrogen bonding in proteins gives rise to two important structures, the (α-helix and β-pleated sheet conformations. Both types of conformation are widely encountered in proteins.

Acids and bases

The biochemical behaviour of many important compounds depends on their acid-base properties. Acids are defined as proton (hydrogen ion) donors and bases as proton acceptors. How readily acids or bases lose or gain protons depends on the chemical nature of the compounds involved. The degree of dissociation of acids in water, for example, ranges from essentially complete dissociation for a strong acid to practically no dissociation for a very weak acid, and any inter-mediate value is possible. It is useful to derive a numerical measure of the strength of an acid, which is the amount of hydrogen ion released when a given amount of acid is dissolved in water. Such an expression, called the acid dissociation constant or K_a, can be written for any acid, HA, that reacts according to the equation

$$\underset{Acid}{HA} \qquad \underset{Conjugate\ base}{H^+ + A^-}$$

$$K_a = \frac{[H^+][A^-]}{[HA]}$$

The square brackets refer to molar concentration, that is, the concentration in moles per liter. For each acid, the quantity K_a has a fixed numerical value at a given temperature. This value is larger for more completely dissociated acids; the greater the K_a the stronger the acid. Strictly speaking, the acid-base reaction we have just written is a protontransfer reaction in which water acts as a base as well-as the solvent.

$$HA(aq) + H_2O(\ell)\ H_3O^+\ (aq) + A^-(aq)$$

acid *Base* *Conjugate*

Conjugate

acid to H_2O base to HA

The notation (aq) refers to solutes in aqueous solution, whereas (ℓ) refers to water in the liquid state. It is well established that there are no "naked protons" (free hydrogen ions) in solution; even the hydronium ion (H_3O^+) is an underestimate of the degree of hydration of hydrogen ion in aqueous solution. All solutes are extensively hydrated in aqueous solution. We will write the short form of equations for acid dissociation in the interests of simplicity, but the role of water should be kept in mind throughout our discussion. The acid-base properties of water play an important part in biological processes because of the central role of water as a solvent. The extent of selfdissociation of water to hydrogen ion and hydroxide ion.

$$H_2O \rightleftharpoons H^+ + OH^-$$

is small, but the fact that it takes place determines important properties of many solutes. Both the hydrogen ion (H^+) and the hydroxide ion (OH^-) are associated with several water molecules, as are all ions in aqueous solution, and the water molecule in the equation is itself part of a cluster of such molecules. It is especially important to have a quantitative estimate of the degree of dissociation of water. We can start with the expression

$$K_a = \frac{[H^+][OH^-]}{[H_2O]}$$

The molar concentration of pure water, $[H_2O]$, is quite large compared with any possible concentrations of solutes and can be considered a constant. (The numerical value is 55.5 M, which can be obtained by dividing the number of grams of water in one liter, 1000 grams, by the molecular weight of water, 18 grams/mole; 1000/18 = 55.5 M.) Thus,

$$K_a = \frac{[H^+][OH^-]}{55.5}$$

$$K_a \times 55.5 = [H^+][OH^-] = K_w$$

A new constant, K_w the ion product constant for water, has just been defined, where the concentration of water has been included in its value. The numerical value of K_w can be determined experimentally by measuring the hydrogen ion concentration of pure water. The hydrogen ion concentration is also equal, by definition, to the hydroxide ion concentration, because water is a monoprotic acid (one that releases a single proton per molecule). At 25°C in pure water,

$$[H^+] = 10^{-7}\ M = [OH^-]$$

Thus, at 25°C the numerical value of K_w is given by the expression

$$K_w = [H^+]\ [OH^-] = (10^{-7})\ (10^{-7}) = 10^{-14}$$

This relationship, which we have derived for pure water, is valid for an aqueous solution, whether neutral, acidic, or basic. The wide range of possible hydrogen ion and hydroxide ion concentrations in aqueous solution makes it desirable to define a quantity for expressing these concentrations more conveniently than by exponential notation. This quantity is called pH and is defined as

$$pH = -\log_{10}[H^+]$$

with the logarithm taken to the base 10. Note that, because of the logarithms involved, a difference of one pH unit implies a tenfold difference in hydrogen ion concentration, $[H^+]$. The pH values of some typical aqueous samples can be determined by a simple calculation.

Monitoring acidity

There is an equation that connects the K_a of any weak acid with the pH of a solution containing both that acid and its conjugate base. This relationship has wide use in biochemical practice, especially where it is necessary to control pH for optimum reaction conditions. Some reactions cannot take place if the pH varies from the optimum value. Important biological macromolecules lose activity at extremes of pH. Also, some drastic physiological consequences can result from pH fluctuations in the body. More information about how pH can be controlled. Tod derive the involved equation, it is first necessary to take the logarithm of both sides of the K_a equation.

$$K_a = \frac{[H^+][A^-]}{[HA]}$$

$$\log K_a = \log [H^+] + \log \frac{[A-]}{[HA]}$$

$$-\log [H^+] = -\log K_a + \log \frac{[A^-]}{[HA]}$$

We then use the definitions of pH and pK_a

$$pH = pK_a + \log \frac{[A^-]}{[HA]}$$

This relationship is known as the Henderson-Hasselbalch equation, and it is of use in predicting the properties of buffer solutions used to control the pH of reaction mixtures. We will be interested in the situation where the concentration of acid, [HA], and the concentration of the conjugate base, $[A^-]$, are equal ($[HA] = [A^-]$). The ratio $[A^-]/[HA]$ is then equal to 1, and the logarithm of 1 to any base is equal to zero. Therefore, when a solution contains equal concentrations of a work acid and its conjugate base the pH of that solution equals the pK_a value of the weak acid.

Titration curves

A titration is an experiment in which measured amounts of base are added to a measured amount of acid. It is convenient and straightforward to follow the course of the reaction with a pH meter. The point in the titration at which the acid is exactly neutralized is called the equivalence point. If the pH is monitored as base is added to a sample of acetic acid in the course of a titration, an inflection point in the titration curve is reached when the pH -equals the pK_a of acetic acid. As we saw in our discussion of the Henderson-Hasselbalch equation, a pH value equal to the pK_a corresponds to a mixture with equal concentrations of the weak acid and its conjugate base—in this case, acetic acid and acetate ion, respectively. The pH at the inflection point is 4.76, which is the pK_a of acetic acid. The inflection point occurs when 0.5 mole of base has been added for each mole of acid present.

Near the inflection point the pH changes very little as more base is added. When 1 mole of base has been added for each mole, the equivalence point is reached, and essentially all the acetic acid has been converted to acetate ion. The relative abundance of acetic acid and acetate ion with increasing additions of NaOH. Notice that the percentage of acetic acid plus the percentage of acetate ion adds up to 100%.

The acid (acetic acid) is progressively converted to its conjugate base (acetate ion) as more NaOH is added and the titration proceeds. It can be helpful to keep track of the percentages of a conjugate acid and base in this way to understand the full significance of the reaction taking place in a titration. Note that these acids are categorized in three groups. The first group is monoprotic acids, which release one hydrogen ion and have a single K_a and pK_a.

The second group is diprotic acids, which release

two hydrogen ions and have two $K_a s$ and two $pK_a s$. The third group is polyprotic acids, which release more than two hydrogen ions. The two examples of polyprotic acids given here, citric acid and phosphoric acid, release three hydrogen ions and have three K_as and three pK_as. Amino acids and pepticles. Here is a way to keep track of protonated and deprotonated forms of acids and their conjugate bases, and this can be particularly useful with diprotic and polyprotic acids.

$$pH < pK_a$$

H^+ on, substance protonated

$$pH > pK_a$$

H^+ off, substance deprotonated

Buffers

A buffer solution consists of a mixture of a weak acid and its conjugate base. Buffer solutions tend to resist a change in pH on addition of moderate amounts of strong acid or base. Let us compare the changes in pH that occur on addition of equal amounts of strong acid or strong base to pure water at pH 7 and to a buffer solution at pH 7. If 1.0 mL of 0.1 M HCI is added to 99.0 mL of pure water, the pH drops drastically. If the same experiment is conducted with 0.1 M NaOH in place of 0.1 M HCl, the pH rises drastically.

The new pH can then be calculated using the Henderson—Hasselbalch equation and the phosphate ion concentrations. The appropriate pK_a is 7.20

$$pH = pK_a + \log \frac{[HPO_4^{2-}]}{[H_2PO_4^-]}$$

$$pH = 7.20 + \log \frac{0.062}{0.101}$$

$$pH = 6.99$$

The new pH is 6.99, a much smaller change than in the unbuffered pure water. Similarly, if 1.0 mL of 0.1 M NaOH is used, the same reaction takes places as in a titration:

$$H_2PO_4^- + OH^- \; HPO_4^{2-} + H_2O$$

Since this buffer is an aqueous solution, it is still true that $K_w = [H^+] [OH^-]$. The increase in hydroxide ion concentration implies that the hydrogen ion concentration decreases and that the pH increases. Use the Henderson-Hasselbalch equation to calculate the new pH and convince yourself that the result is pH = 7.01, again a much smaller change in pH than took place in pure water.

Many biological reactions will not take place unless the pH remains within fairly narrow limits, and as a result buffers have great practical importance in the biochemistry laboratory. A consideration of titration curves can give insight into the origin of this property. The pH of a sample being titrated changes very little the vicinity of the inflection point of a titration curve. Also, at the inflection point half the amount of acid originally present has been converted to the conjugate base.

The second stage of ionization of phosphoric acid, $H_2PO_4^- \; H^+ + HPO_4^{2-}$ was the basis of the buffer just used as an example. The pH at the inflection point of the fitration is 7.20, a value numerically to the pK_a of the dihydrogen phosphate ion. At this pH the solution contains equal concentrations of the dihydrogen phosphate and monohydrogen phosphate ions, the acid and base forms.

A buffer solution can maintain the pH at a relatively constant value because of the presence of appreciable amounts of both the acid and its conjugate base. This condition is met at pH values at or near the pK_a of the acid. It OH is added, an appreciable amount of the acid form of the buffer is present in solution to react with the added base. If H^+ is added, there is also an appreciable amount of the basic form of the buffer to react with the added acid. The $H_2PO_4^-/HPO_4^{2-}$ pair is suitable as a buffer near PH 7.2. and the CH_3COOH/CH_3COO^- pair near pH 4.76. At PH values below the pK_a the acid form predominates, and at PH values above the pK_a the basic form predominates. An acid-to-base ratio of 10: 1 corresponds to a PH lower than the pK_a by about 1 PH unit, because each unit change in the PH scale corresponds to a tenfold change in the hydrogen ion concentration.

An acid-to base of 1:10 corresponds to a PH about 1 unit higher than the pK_a. The plateau region in a titration curve, where the PH does not change rapidly, covers a PH range extending approximately 1 PH unit on each side of the pK_a. Thus, there is a range of about 2 PH units in which the buffer is effective. The condition that a buffer contains appreciable amounts of both a weak acid and its conjugate base applies both to the ratio of the two forms and to the absolute amount of each present in a given solution. If a buffer solution contained a suitable ratio of acid to base, but very low concentrations of both, it would take very little added acid to use up all of the base form, or vice versa. A buffer solution with low concentrations of both the acid and base forms is said to have a low buffering capacity. A buffer that contains greater amounts of both acid and base has a higher buffering capacity. The following essay describes some of the considerations that go into the choice of a suitable buffer for a given application.

Buffer systems of physiological importance

Buffer systems in living organisms and in the laboratory are based on many types of compounds. Since physiological PH in most organisms stays around 7, it might be expected that the phosphate buffer system would be widely used in living organisms. This is the case in which phosphate ion concentrations are high enough for the buffer to be effective, as in most intracellular fluids. The $H_2PO^-_4/HPO_4^{2-}$ pair is the principal buffer in cells. In blood, phosphate ion levels are inadequate for buffering, and a different system operates. The buffering system in blood is based on the dissociation of carbonic acid (H_2CO_3)

$$H_2CO_3 \rightleftharpoons H^+ + HCO_3^-$$

where the pK_a of H_2CO_3 is 6.37. The pH of human blood, 7.4, is near the end of the buffer range of this system, but another factor enters into the situation. Carbon dioxide can dissolve in water and in water-based fluids, such as blood. The dissolved carbon dioxide forms carbonic acid, which in turn reacts to produce bicarbonate ion, HCO_3^-

$$CO_2(g) \rightleftharpoons CO_2(aq)$$

$$CO_2(aq) + H_2O\ (\ell) \rightleftharpoons H_2CO_3(aq)$$

$$H_2CO_3(aq) \rightleftharpoons H^+(aq) + HCO_3^-(aq)$$

$$\text{Net equation: } CO_2(g) + H_2O(\ell) \rightleftharpoons H^+(aq) + HCO_3^-(aq)$$

At the pH of blood, which is about 1 unit higher than the pK_a of carbonic acid, most of the dissolved CO_2 is present as HCO_3^-. The CO_2 being transported to the lungs to be expired takes the form of bicarbonate ion. There is a direct relationship between the pH of the blood and the pressure of carbon dioxide gas in the lungs. The properties of hemoglobin, the oxygen carrying protein in the blood, also enter into the situation. The phosphate buffer system is common in the laboratory (in vitro, outside the living body) as well as in living organisms (in vivo).

The buffer system based on tris (hydroxymethyl) aminomethane (called TRIS) is also widely used in vitro. Other buffers that have come into wide use more recently are zwitterions, compounds that each have both a positive charge and a negative charge. Zwitterions are usually considered less likely to interfere with biochemical reactions than some of the earlier buffers.

Chapter 3

Enzymes

Enzymes, like all catalysts, speed up reactions. They cannot alter the equilibrium constant or the free energy change. The reaction rate depends on the free energy of activation (activation energy) ($\Delta G^{o\ddagger}$), the energy input required to initiate the reaction. The activation energy for an uncatalysed reaction is higher than that for a catalysed reaction; in other words, an uncatalysed reaction requires more energy to get started, and that is why its rate is slower than that of a catalysed reaction.

The energy that must be supplied to start the reaction (which then proceeds with a release of energy)—the activation energy—is conceptually similar to the act of pushing an object to the top of a hill so that it can then slide down the other side. The activation energy directly affects the rate of reaction, and the presence of a catalyst speeds up a reaction by changing the mechanism and thus lowering the activation energy. An important effect of a catalyst on a chemical reaction is apparent from a comparison of the activation energy profiles of the same reaction, catalysed and uncatalysed.

The standard free energy change for the reaction, ΔG°, remains unchanged on addition of a catalyst, but the activation energy, ΔG is lowered. The biochemical reaction in which hydrogen peroxide (H_2O_2) is converted to water and oxygen provides an example of the effect of catalysts on activation energy.

$$2H2O_2 \rightarrow 2H_2O + O_2$$

Activation energy of this reaction is lowered if the reaction is allowed to proceed on platinum surfaces, but it is lowered even more by the enzyme catalase. Raising the temperature of a reaction mixture increases the energy available to the reactants to reach the transition state. Consequently, the rate of a chemical reaction increases with temperature, and one might be tempted to assume that this is universally true for biochemical reactions. In fact, increase of reaction rate with temperature occurs only to a limited extent with biochemical reactions. It is helpful to raise the temperature at first, but eventually a point is reached at which heat denaturation of the enzyme plays a part.

Enzyme kinetics

Rate of a chemical reaction is expressed in terms of a change in the concentration of a reactant or of a product in a given time interval. Any convenient experimental method can be used to monitor changes in concentration. In a reaction of the form A + B → P, where A and B are reactants and P is the product, the rate of reaction can be expressed in terms of either the rate of disappearance of one of the reactants or the rate of appearance of product. The rate of disappearance of A is $-\Delta[A]/\Delta t$, where Δ symbolizes change, [A] is the concentration of A in moles/liter^{-1}, and t is time. Likewise, the rate of disappearance of B is $\Delta [B]/\Delta t$, and the rate of appearance of P is $\Delta [P]/\Delta t$. The rate of the reaction can be expressed in terms of any of these changes, because the rates of appearance of product and disappearance of reactant are related by the stoichiometric equation for the reaction.

$$\text{Rate} = -\frac{\Delta[\text{A}]}{\Delta t} = -\frac{\Delta[\text{B}]}{\Delta t} = \frac{\Delta[\text{P}]}{\Delta t}$$

It has been established that the rate of a reaction at a given time is proportional to the product of the concentrations of the reactants raised to the appropriate powers.

$$\text{Rate} \propto [\text{A}]^f [\text{B}]^g$$

or,

$$\text{Rate} = k[\text{A}]^f [\text{B}]^g$$

Where k is a proportionality constant called the rate constant. The exponents f and g must be determined experimentally. They are not necessarily equal to the coefficients of the balanced equation, but frequently they are. The square brackets, as usual, denote molar concentration. When the exponents in the rate equation have been determined experimentally, a mechanism for the reaction—a description of the detailed steps along the path between reactants and products—can be proposed. The exponents in the rate equation are usually small whole numbers such as 1 or 2. (There are also some cases in which the exponent 0 occurs.) The values of the exponents are related to the number of molecules involved in the detailed steps that constitute the mechanism. The overall order of a reaction is the sum of all the exponents. If, for example, the rate of a reaction A $\rightarrow$ P is given by the rate equation

$$\text{Rate} = k[\text{A}]^1$$

Where k is the rate constant and the exponent for the concentration of A is 1, then the reaction is first order with respect to A, and first order overall. The rate of radioactive decay of the widely used tracer isotope phosphorus 32 depends only on the concentration of ^{32}p present; here we have an example of a first-order reaction. Only the ^{32}P atoms are involved in the mechanism of the radioactive decay that, as an equation, takes the form

$$^{32}\text{P} \rightarrow \text{decay products}$$

$$\text{Rate} = k[^{32}\text{P}]^1 = k[^{32}\text{P}]$$

If the rate of a reaction A + B $\rightarrow$ C + D is given by

$$\text{Rate} = k[\text{A}]^1\ [\text{B}]^1$$

(where k is the rate constant, the exponent for the concentration of A is 1, and the exponent for the concentration of B is 1) then the reaction is said to be first order with respect to A, first order with respect to B, and second order overall. In the reaction of methyl bromide (CH_3Br) with hydroxide ion (OH-) to give methyl alcohol (CH_3OH) and bromide ion (Br^-), the rate of reaction depends on the concentrations of both reactants.

$$CH_3Br + OH^- \rightarrow CH_3OH + Br^-$$

$$\text{Rate} = k[CH_3Br]^1\,[OH^-]^1 = k[CH_3Br]\ [OH^-]$$

Both the methyl bromide and the hydroxide ion take part in the reaction mechanism. The reaction of methyl bromide with hydroxide ion is first order with respect to CH_3Br, first order with respect to OH^-, and second order overall. Many common reactions are first or second order. Once the order of the reaction is determined experimentally, conclusions can be drawn about the mechanism of a reaction. The possibility exists that exponents in a rate equation may be equal to zero, with the rate for a reaction A $\rightarrow$ B given by the equation

$$\text{Rate} = k[\text{A}]^0 = kc$$

Such a reaction is called zero order, and its rate, which is constant, depends not on concentrations of reactants, but on other factors such as the presence of catalysts. Enzyme-catalysed reactions can exhibit zero-order kinetics when the concentrations of reactants are so high that the enzyme is completely saturated with reactant molecules.

Formation of enzyme-substrate complex

In an enzyme-catalysed reaction, the enzyme binds to the substrate to form a complex. The formation of the complex leads, in turn, to formation of the transition-state species, which then forms the product. The nature of transition states in enzymatic reactions is a large field of research in itself, but some general statements can be made on the subject. A substrate binds, usually by noncovalent interactions, to a small portion of the enzyme called the active site, frequently situated in a cleft or crevice in the protein and consisting of certain amino acids that are essential for enzymatic activity.

The catalysed reaction takes place at the active site, usually in several steps. The first step is the binding of substrate to the enzyme, which occurs because of highly specific interactions between the substrate and the side chains of the amino acids making up the active site. Two important models have been developed to describe the binding process. The first, the lock-and-key model, assumes a high degree of similarity between the shape of the substrate and the geometry of the binding site on the enzyme. The substrate binds to a site whose shape complements its own, like a key in a lock or the right piece in a three-dimensional jigsaw puzzle.

This model is now largely of historical interest because it does not take into account an important property of proteins, namely, their conformational flexibility. The second model takes into account the fact that proteins have some three-dimensional flexibility. According to this induced-fit model, the binding of the substrate induces a conformational change in the enzyme that results in a complementary fit once the substrate is bound. The binding site has a different three-dimensional shape before the substrate is bound.

In the transition state, the substrate is bound close to atoms with which it is to react. Furthermore, the substrate is placed in the correct orientation with respect to those atoms. Both effects, proximity and orientation, speed up the reaction. As bonds are broken and new bonds are formed, the substrate is transformed into product. The product is released from the enzyme, which can then catalyse the reaction of more substrate to form more product.

Each enzyme has its own unique mode of catalysis, which is not surprising in view of enzymes' great specificity. Even so, there are some general modes of catalysis in enzymatic reactions. Chymotrypsin is an enzyme that catalyses the hydrolysis of peptide bonds, with some specificity for residues containing aromatic side chains. Chymotrypsin cleaves peptide bonds at other sites as well, such as leucine, histidine, and glutamine, but with a lower frequency than at aromatic amino acid residues. It also catalyses the hydrolysis of ester bonds. Although ester hydrolysis is not important to the physiological role of chymotrypsin in the digestion of proteins, it is a convenient model system for investigating the enzyme's catalysis of hydrolysis reactions.

A usual laboratory procedure is to use β-nitrophenyl esters as the substrate and to monitor the progress of the reaction by the appearance of a yellow colour in the reaction mixture due to the production of *p*-nitrophenolate ion. In a typical reaction in which a *p*-nitrophenyl ester is hydrolyzed by chymotrypsin, the experimental rate of the reaction depends on the concentration of the substrate, in this case the *p*-nitrophenyl ester. At low substrate concentrations the rate of reaction increases as more substrate is added. At higher substrate concentrations the rate of the reaction changes very little with addition of more substrate, and a maximum rate is reached. When these results are presented in a graph, the curve is hyperbolic.

Another example for enzyme-catalyzed reaction is the one catalyzed by enzyme aspartate transcarbamoylase (ATCase). This reaction is the first step in a pathway leading to the formation of cytidine triphosphate (CTP) and uridine triphosphate (UTP), which are ultimately needed for the biosynthesis of RNA and DNA. In this reaction, carbamoyl phosphate reacts with aspartate to produce carbamoyl aspartate and phosphate ion.

$$\text{Carbamoyl phosphate} + \text{aspartate} - \text{carbamoyl aspartate} + HPO_4^{2-}$$

Rate of this reaction also depends on substrate concentration, in this case, the concentration of aspartate. Experimental results show that once again the rate of the reaction depends on substrate concentration at low and moderate concentrations, and once again a maximum rate is reached at high substrate concentrations. There is, however, one very important difference. For this reaction, a graph that shows the dependence of reaction rate on substrate concentration has a sigmoidal rather than hyperbolic shape.

The results of experiments on the reaction kinetics of chymotrypsin and aspartate transcarbamoylase are representative of experimental results obtained with many enzymes. The overall kinetic behaviour of many enzymes resembles that of chymotrypsin, and other enzymes behave similarly to aspartate transcarbamoylase.

Comparison between the kinetic behaviours of chymotrypsin and ATCase is reminiscent of the relationship between the oxygen-binding behaviours of myoglobin and hemoglobin. ATCase and hemoglobin are allosteric proteins; chymotrypsin and myoglobin are not. The differences in behaviour between allosteric and nonallosteric proteins can be understood in terms of models based on structural differences between the two kinds of proteins.

Michaelis-Menten approach to enzyme kinetics

In 1913, Leonor Michaelis and Maud Menten devised a useful model for the kinetics of enzyme-catalyzed reactions. Theirs is still the basic model for nonallosteric enzymes. It is widely used even though it has undergone many modifications. A typical reaction might be the conversion of some substrate, S, to a product, P. The stoichiometric equation for the reaction is

$$S \rightarrow P$$

Mechanism for an enzyme-catalyzed reaction can be summarized in the form of

$$E + S \xrightarrow{k_1} ES \xrightarrow{k_2} E + P$$

The product is not converted to substrate to any appreciable extent. In this equation k_1 is the rate constant for the formation of the enzyme-substrate complex, ES, from the enzyme, E, and the substrate, S; k_{-1} is the rate constant for the reverse reaction, dissociation of the ES complex to free enzyme and substrate; and k_2 is the rate constant for the conversion of the ES complex to product P and the subsequent release of product from the enzyme. The enzyme appears explicitly in the mechanism, and the concentrations of both free enzyme, E, and enzyme-substrate complex, ES, therefore appear in the rate equations. It is a characteristic of catalysts that they are regenerated at the end of the reaction, and this is true of enzymes.

While measurimg the rate of an enzymatic reaction at varying substrate concentrations, we see that the rate depends on the substrate concentration, [S]. We normally measure the initial rate of the reaction (the rate measured immediately after the enzyme and substrate are mixed) so that we can be certain that the product is not converted to substrate to any appreciable extent.

In the lower region of the curve (at low levels of substrate), the reaction is first order, implying that the velocity, V, depends on substrate concentration [S]. In the upper portion of the curve (at higher levels of substrate), the reaction is zero order; the rate is independent of concentration. The active sites of all enzyme molecules are saturated. The reaction proceeds at its maximum velocity, written V_{max}. The substrate concentration at which the reaction proceeds at one -half its maximum velocity has a special significance. It is given the symbol K_M, which can be considered an inverse measure of the affinity of the enzyme for the substrate. The lower the K_M, the higher the affinity. Let us examine the mathematical relationships among the quantities [E], [S], V_{max}, and K_M. The general mechanism of the enzyme-catalyzed reaction involves binding of the enzyme, E, to the substrate to form a complex, ES, which then forms the product.

Negative sign in the term $\Delta[ES]/\Delta t$ means that the concentration of the complex decreases as the complex breaks down. The term k_{-1} is the rate constant for the dissociation of complex to regenerate enzyme and substrate, and k_2 is the rate constant for the reaction of the complex to give product and enzyme. Enzymes are capable of processing the substrate very efficiently, and a steady state is soon reached in which the rate of formation of the enzyme-substrate complex equals the rate of its breakdown. Very little complex is present, and it turns over rapidly, but its concentration stays the same with time. According to the steady-state theory, then, the rate of formation of the enzyme-substrate complex equals the rate of its breakdown,

$$\frac{\Delta[\text{ES}]}{\Delta t} = -\frac{\Delta[\text{ES}]}{\Delta t}$$

and

$$k_1[\text{E}]\,[\text{S}] = k^{-1}\,[\text{ES}] + k_2[\text{ES}]$$

To solve for the concentration of the complex ES, it is necessary to know the concentration of the other species involved in the reaction. The initial concentration of substrate is a known experimental condition and does not change significantly during the initial stages of the reaction. The substrate concentration is much greater than the enzyme concentration. The total concentration of the enzyme, $[\text{E}]_T$, is also known, but a large proportion of it may be involved in the complex.

The concentration of free enzyme, [E], is the difference between $[\text{E}]_T$, the total concentration and [ES], which can be written as an equation:

$$[\text{E}] = [\text{E}]_T - [\text{ES}]$$

Substituting for the concentration of free enzyme, [E], in Equation

$$k_1\left([\text{E}]_T - [\text{ES}]\right)[\text{S}] = k_{-1}[\text{ES}] + k_2[\text{ES}]$$

Collecting all the rate constants for the individual reactions,

$$\frac{\left([\text{E}]_T - [\text{ES}]\right)[\text{S}]}{[\text{ES}]} = \frac{k_{-1} + k_2}{k_1} = K_M$$

Where K_M is called the Michaelis constant. It is now

possible to solve Equation for the concentration of enzyme-substrate complex, [ES],

$$\frac{([E]_T[S]-[ES][S])}{[ES]} = K_M$$

$$[E]_T[S]-[ES][S] = K_M[ES]$$

$$[E]_T[S] = [ES](K_M+[S])$$

or

$$[ES] = \frac{[E]_T[S]}{K_M+[S]}$$

In the initial stages of the reaction so little product is present that no reverse reaction of product to complex need be considered. It is the initial rate that is usually determined in enzymatic reactions, and this rate depends on the rate of breakdown of the enzyme-substrate complex into product and enzyme.

In the Michaelis-Menten model, the initial rate (V_{init}; in some texts the notation is V_0) of formation of product depends only on the rate of the breakdown of the ES complex. If the substrate concentration is so high that the enzyme is completely saturated with substrate ($[ES] = [E]_T$), the reaction proceeds at its maximum possible rate (V_{max}), and, substituting $[E]_T$ for [ES] in Equation

$$V_{init} = V_{max} = k_2[E]_T$$

Original concentration of enzyme is a constant, i.e.,

$$V_{max} = \text{constant}$$

Expression for V_{max} resembles that for a zero-order reaction given in Equation

$$\text{Rate} = k[\text{A}]^0 = k$$

Concentration of substrate, [A], appears in Equation rather than the concentration of enzyme, [E]. When the enzyme is saturated with substrate, zero-order kinetics with respect to substrate are observed; thus, it is necessary to use comparatively low concentrations of substrate to determine initial rates of enzymatic reactions.

At higher substrate concentrations, when the enzyme is saturated, the constant reaction rate characteristic of zero-order kinetics is observed.

$$V = \frac{V_{max}[\text{S}]}{K_M + [\text{S}]}$$

When experimental conditions are adjusted so that $[\text{S}] = K_M$

$$V = \frac{V_{max}[\text{S}]}{[\text{S}] + [\text{S}]}$$

and

$$V = \frac{V_{max}}{2}$$

When the rate of the reaction is one-half its maximum value, the substrate concentration is equal to the Michaelis constant, and this fact is the basis of the graphical determination of K_M. A curve that describes the rate of a nonallosteric enzymatic reaction is hyperbolic. It is quite difficult to determine a single point at which the rate levels off. This, in turn, makes it difficult to determine the V_{max} and thus the K_M of the enzyme. It is considerably easier to work with a straight line than a curve. One can transform the equation, for a hyperbola.

$$V = \frac{V_{max}[S]}{K_M + [S]}$$

Into an equation for a straight line by taking the reciprocals of both sides:

$$\frac{1}{V} = \frac{K_M + [S]}{V_{max} + [S]}$$

$$\frac{1}{V} = \frac{K_M}{V_{max} + [S]} + \frac{[S]}{V_{max} + [S]}$$

$$\frac{1}{V} = \frac{K_M}{V_{max}} \cdot \frac{1}{[S]} + \frac{1}{V_{max}}$$

The equation now has the form of a straight line, $y = mx + b$, where $1/V$ takes the place of the y coordinate and $1/[S]$ takes the place of the x coordinate. The slope of the line, m, is K_M/V_{max}, and the intercept, b, is $1/V_{max}$.

When the rate of a reaction, V, is equal to one-half the maximum rate possible, $V = V_{max}/2$, then $K_M = [S]$. One interpretation of the Michaelis constant K_M is that it equals the concentration of substrate at which 50% of the enzyme active sites are occupied by substrate. The Michaelis constant has the units of concentration. Another interpretation of K_M relies on the assumptions of the original Michaelis-Menten model of enzyme kinetics. From Equation, we know that,

$$E + S \xrightarrow{k_1} ES \xrightarrow{k_2} E + P$$

Where, k_1 is the rate constant for the formation of the enzyme-substrate complex, ES, from the enzyme and substrate; k_{-1} is the rate constant for the reverse reaction, dissociation of the ES complex to free enzyme and substrate; and k_2 is the rate constant for the formation of product P and the subsequent release of product from the enzyme. Also Equation shows

$$K_M = \frac{k_{-1} + k_2}{k_1}$$

In case, where the reaction E + S $\rightarrow$ ES takes place more frequently than E + S $\rightarrow$ E + P. In kinetic terms, this means that the dissociation rate constant k_{-1} is greater than the rate constant for the formation of product, k_2. If k_{-1} is much larger than $k_2(k_{-1} >> k_2)$, as is assumed by the steady-state model, then approximately,

$$K_M = \frac{k_{-1}}{k_1}$$

It is informative to compare the expression for the Michaelis constant with the equilibrium constant expression for the dissociation of the ES complex,

$$E + S \xrightarrow{k_1} ES$$

The equilibrium constant expression is

$$K_{eq} = \frac{[E][S]}{[ES]} = \frac{k_{-1}}{k_1}$$

This expression is the same as that for K_M and makes the point that when the assumption that $k_{-1} >> k_2$ is valid, K_M is simply the dissociation constant for the ES complex. The K_M is a measure of how tightly the substrate is bound to the enzyme. The greater the value of the K_M, the less tightly the substrate is bound to the enzyme. The V_{max} is related to the turnover number of an enzyme, a quantity equal to the catalytic constant, k_{cat}.

$$\frac{V_{max}}{[E_T]} = \text{turnover number} = k_{cat}$$

The turnover number is the number of moles of substrate that react to form product per mole of enzyme per unit time. This statement assumes that the enzyme is fully saturated with substrate and, thus, that the reaction is proceeding at the maximum rate. Turnover numbers are a particularly dramatic illustration of the efficiency of enzymatic catalysis.

Inhibition of enzymatic reactions

Inhibitor is a substance that interferes with the action of an enzyme and slows the rate of a reaction. A good deal of information about enzymatic reactions can be obtained by observing the changes in the reaction caused by the presence of inhibitors. There are two ways in which inhibitors can affect an enzymatic reaction. A reversible inhibitor can bind to the enzyme and subsequently be released, leaving the enzyme in its original condition. An irreversible inhibitor reacts with the enzyme to produce a protein that is not enzymatically active, and from which the original enzyme cannot be regenerated. Two major classes of reversible inhibitors can be distinguished on the basis of the sites on the enzyme to which they bind.

One class consists of compounds very similar in structure to the substrate. In this case the inhibitor can bind to the active site and block the substrate's access to it. This mode of action is called competitive inhibition because the inhibitor competes with the substrate for the active site on the enzyme.

The other major class of reversible inhibitors includes any inhibitor that binds to the enzyme at a site other than the active site and, as a result of binding, causes a change in the structure of the enzyme, especially around the active site. The substrate may still be able to bind to the active site, but the enzyme cannot catalyse the reaction as efficiently as it could in the absence of the inhibitor. This mode of action is called noncompetitive inhibition. The two kinds of inhibition can be distinguished from one another in the laboratory. The reaction is carried out in the presence of inhibitor at several substrate concentrations, and the rates obtained are compared with those of the uninhibited reaction. The differences in the Lineweaver-Burk plots for the inhibited and uninhibited reactions provide the basis for the comparison.

In the presence of a competitive inhibitor, the slope of the Lineweaver-Burk plot changes, but the y intercept does not. The V_{max} is unchanged, but the K_M increases. More substrate is needed to get to a given rate, including V_{max} in the presence of inhibitor rather than in its absence. This point specifically applies to the specific value $V_{max}/2$ (recall that at $V_{max}/2$, the substrate concentration, [S], equals K_M). Competitive inhibition can be overcome by a sufficiently high substrate concentration. In the presence of a competitive inhibitor the equation for an enzymatic reaction becomes

$$EI \xrightarrow{I} E \xrightarrow{S} ES \longrightarrow E + P$$

Where EI is the enzyme-inhibitor complex. The dissociation constant for the enzyme-inhibitor complex can be written:

$$EI \longrightarrow E + I$$

In the presence of inhibitor the value of K_M increases by the factor

$$1 + \frac{[1]}{K_1}$$

If we substitute K_M (1 + [I] /K_I]K_M in Equation, we obtain

$$\frac{1}{V} = \frac{K_M}{V_{max}} \cdot \frac{1}{[S]} + \frac{1}{V_{max}}$$

$$\frac{1}{V} = \frac{K_M}{V_{max}}\left(1 + \frac{[I]}{K_I}\right) \cdot \frac{1}{[S]} + \frac{1}{V_{max}}$$

$$y = m \,.\, x + b$$

The term $1/V$ itakes the place of the y coordinate, and the term I/[S] takes the place of the x coordinate, as was the case in Equation. The intercept I/V_{max} the b term in the equation for a straight line, has not changed from the earlier equation, but the slope K_M/

V_{max} in Equation has increased by the factor $(1 + [I]/K_I)$. The slope, the *m* term in the equation for a straight line, is now

$$\frac{K_M}{V_{max}}\left(1+\frac{[I]}{K_I}\right)$$

Accounting for the changes in the slope of the Lineweaver-Burk plot. Note that the *y* intercept does not change. This algebraic treatment of competitive inhibition agrees with experimental results, validating the model, just as experimental results validate the underlying Michaelis-Menten model for enzyme action.

The kinetic results of noncompetitive inhibition differ from those of competitive inhibition. The Lineweaver-Burk plots for a reaction in the presence and absence of a noncompetitive inhibitor show that both the slope and the *y* axis intercept change for the inhibited reaction, without changing the *x* axis intercept. The value of V_{max} decreases, but that of K_M remains the same; the inhibitor does not interfere with the binding of substrate to the active site. Increasing the substrate concentration cannot overcome noncompetitive inhibition. The reaction pathway has become considerably more complicated, and several equilibrium have to be considered.

$$E \xrightarrow{+S} ES \longrightarrow E + P$$

$$+I \qquad\qquad +I$$

$$EI \xrightarrow{+S} ESI$$

In the presence of a noncompetitive inhibitor, I, the maximum velocity of the reaction, V^{I}_{max}, has the form

$$V^{I}_{max} = \frac{V_{max}}{1 + [I]/K_I}$$

Here K_I is the dissociation constant for the enzyme-inhibitor complex EI. The maximum rate V_{max} appears in the expressions for both the slope and the intercept in the equation for the Lineweaver-Burk plot:

$$\frac{1}{V} = \frac{K_M}{V_{max}} \cdot \frac{1}{[S]} + \frac{1}{V_{max}}$$

$$y = m.x + b$$

In noncompetitive inhibition, we replace the term V_{max} with the expression for V_{max} to obtain

$$\frac{1}{V} = \frac{K_M}{V_{max}}\left(1 + \frac{[I]}{K_I}\right) \cdot \frac{1}{[S]} + \frac{1}{V_{max}}\left(1 + \frac{[I]}{K_I}\right)$$

The expressions for both the slope and the intercept in the equation for a Lineweaver-Burk plot of an uninhibited reaction have been replaced by more complicated expressions in the equation that describes noncompetitive inhibition. This interpretation is borne out by the observed results. Sucrose is hydrolyzed to glucose and fructose in a classic experiment in kinetics. The reaction is catalyzed by the enzyme invertase. Using the following data, determine by the Lineweaver-Burk method whether the inhibition of this reaction by 2 M urea is competitive or noncompetitive.

Behaviour of many enzymes can be described by the Michaelis-Menten model, but allosteric enzymes behave very differently. There are similarities between the reaction kinetics of an enzyme such as chymotrypsin, which does not display allosteric behaviour, and the binding of oxygen by myoglobin, also an example of nonallosteric behaviour. The analogy extends to show the similarity in the kinetic behaviour of an allosteric enzyme such as aspartate transcarbamoylase (ATCase) and the binding of oxygen by hemoglobin. Both ATCase and hemoglobin are allosteric proteins; the behaviours of both exhibit cooperative effects caused by subtle changes in quaternary structure. In addition to displaying cooperative kinetics, allosteric enzymes have a different response to the presence of inhibitors from that of nonallosteric enzymes, as characterized by the Michaelis-Menten model.

ATCase catalyzes the first step in a series of reactions with the end product cytidine triphosphate (CTP). CTP is an inhibitor of ATCase, the enzyme that catalyzes the first reaction in the pathway. This behaviour is an example of feedback inhibition, in which the end product of the sequence of reactions inhibits the first reaction in the series. Feedback inhibition is an efficient control mechanism because the entire series of reactions can be shut down when an excess of the final product exists; intermediates in the pathway are not formed.

Feedback inhibition is a general feature of metabolism and is not confined to allosteric enzymes. However, the observed kinetics of the ATCase reaction, including the mode of inhibition, are typical of allosteric enzymes. When ATCase catalyzes the condensation of aspartate and carbamoyl phosphate to form carbamoyl aspartate, the graphical representation of the rate as a function of increasing substrate concentration is a sigmoidal curve rather than the hyperbola obtained with nonallosteric enzymes. The sigmoidal curve is indicative of the cooperative behaviour of allosteric enzymes.

At high substrate concentrations, the same maximal rate (V_{max}) is observed in the presence and absence of inhibitor. Since, in the Michaelis-Menten scheme, the V_{max} changes when a reaction takes place in the presence of a noncompetitive inhibitor, noncompetitive inhibition cannot be the case here. The same Michaelis-Menten model associates this sort of behaviour with competitive inhibition, but that part of the model still does not provide a reasonable picture. Competitive inhibitors bind to the same site as the substrate because they are very similar in structure. The CTP molecule is very different in structure from the substrate, aspartate. It is much more likely that CTP is bound to a different site on the ATCase molecule. There is, in fact, experimental evidence for two different binding sites: The enzyme can be modified so that CTP cannot bind, but the binding of aspartate is not affected by this modification. The situation becomes curiouser and curiouser when the ATCase reaction takes place not in the presence of CTP, a pyrimidine nucleoside triphosphate, but in the presence of ATP, a purine nucleoside triphosphate.

Structural similarities between CTP and ATP are apparent, but ATP is not a product of the pathway that includes the reaction of ATCase and that produces CTP. Both ATP and CTP are needed for the synthesis of RNA and DNA. The relative proportions of ATP and CTP are specified by the needs of the organism. If there is not enough CTP compared to ATP, the enzyme requires a signal to produce more.

In the presence of ATP, the rate of the enzymatic reaction is increased at lower levels of aspartate and the shape of the rate curve becomes less sigmoidal and more hyperbolic. In other words, there is less cooperativity in the reaction. The binding site for ATP on the enzyme molecule is the same as that for CTP, which is not surprising in view of their structural similarity, but ATP is an activator rather than an inhibitor like CTP. When CTP is in short supply in an organism, the ATCase reaction is not inhibited, and the binding of ATP increases the activity of the enzyme still more.

The key to allosteric behaviour, including cooperativity and modifications of cooperativity, is the existence of multiple forms for the quaternary structures of allosteric proteins. The word "allosteric" is derived from allo, "other," and steric, "shape," referring to the fact that the possible conformations affect the behaviour of the protein. The binding of substrates, inhibitors, and activators changes the quaternary structure of allosteric proteins, and the changes in structure are reflected in behaviour. A substance that modifies the quaternary structure, and thus the behaviour, of an allosteric protein by binding to it is called an allosteric effector. The term "effector" can apply to substrates, inhibitors, or activators. Several models for the behaviour of allosteric enzymes have been proposed, and it is worthwhile to compare them.

Homotropic effects are allosteric interactions that occur when several identical molecules are bound to a protein. The binding of substrate molecules to different sites on an enzyme, such as the binding of aspartate to ATCase, is an example of a homotropic effect.

Heterotropic effects are allosteric interactions that occur when different substances are bound to the protein. In the ATCase reaction, inhibition by CTP and activation by ATP are both heterotropic effects.

The two principal models for the behaviour of allosteric enzymes are the concerted model and the sequential model. They were proposed in 1965 and 1966, respectively, and both are in current use as bases for interpreting experimental results. The concerted model has the advantage of comparative simplicity, and it describes the behaviour of some enzyme systems very well. The sequential model sacrifices a certain amount of simplicity for a more realistic picture of the structure and behaviour of proteins.

Jacques Monod, Jeffries Wyman, and Jean-Pierre Changeux proposed the concerted model for the behaviour of allosteric proteins in a paper that has become a classic in the biochemical literature. In this picture the protein has two conformations, the R (relaxed) conformation, which binds substrate tightly, and the T (tight, also called taut) conformation, which binds substrate less tightly. The distinguishing feature of this model is that the conformations of all subunits change simultaneously. Both subunits change conformation from the inactive T conformation to the active R conformation at the same time; that is, a concerted change of conformation occurs. The binding of the first molecule of substrate to one subunit facilitates the binding of the second substrate molecule to the other subunit. This is exactly what is meant by cooperative binding. In the absence of substrate, the enzyme exists mainly in the T form, in equilibrium with small amounts of the R form. The presence of substrate shifts the equilibrium to produce more of the R form.

In concerted model, the effects of inhibitors and activators can also be considered in terms of shifting the equilibrium between the T and R forms of the enzyme. The binding of inhibitors to allosteric enzymes is cooperative; allosteric inhibitors bind to and stabilize the T form of the enzyme. The binding of activators to allosteric enzymes is cooperative as well; allosteric: activators bind to and stabilize the R form of the enzyme.

When an activator (A) is present, the cooperative binding of A shifts the equilibrium between the T and R forms; with the R form favoured. As a result, there is less need for substrate (S) to shift the equilibrium in favour of the R form and less need for cooperativity in the binding of S. When an inhibitor (I) is present, the cooperative binding of I also shifts the equilibrium between the T and R forms, but this time the T form is favoured. More substrate is needed to shift the T-to-R equilibrium in favour of the R form. A greater degree of cooperativity is needed in the binding of S.

Daniel Koshland introduced the direct sequential model of allosteric behaviour. The distinguishing feature of this model is that the binding of substrate induces the conformational change from the T form to the R form—the type of behaviour postulated by the induced-fit theory of substrate binding. The change from T to R in one subunit favours the same change in the other subunits. A conformational change from T to R in one subunit makes the same conformational change easier in another subunit, and this is the form in which cooperative binding is expressed in this model. In the sequential model, the binding of activators and inhibitors also takes place by the induced-fit mechanism. The conformational change that begins with binding of inhibitor or activator to one subunit affects the conformations of other subunits. The net result is to favour the R state when activator is present and the T form when inhibitor (1) is present. Binding of I to one subunit causes a conformational change so that the T form is even less likely to bind substrate than before. This conformational change is passed along to other subunits, making them also more likely to bind inhibitor and less likely to bind substrate. This is an example of cooperative behaviour that leads to more inhibition of the enzyme. Likewise, binding of an activator causes a conformational change that favours substrate binding, and this effect is passed from one subunit to another.

Sequential model for the binding of effectors of all types, including substrates, to allosteric enzymes has a unique feature. Another possible type of change in the enzyme behaviour brought about by the binding of the first molecule. The conformational changes thus induced can make the enzyme less likely to bind more molecules of the same type. This phenomenon, called negative cooperativity, has been observed in a few enzymes. One is tyrosyl tRNA synthetase, which plays a role in protein synthesis. In the reaction catalyzed by this enzyme, the amino acid tyrosine forms a covalent bond to a molecule of transfer RNA (tRNA). In subsequent steps the tyrosine is passed along to its place in the sequence of the growing protein. The tyrosyl tRNA synthetase consists of two subunits. Binding of the first molecule of substrate to one of the subunits inhibits binding of a second molecule to the other subunit. The sequential model has successfully accounted for the negative cooperativity observed in the behaviour of tyrosyl tRNA synthetase.

Zymogens

Zymogen is an inactive precursor of an enzyme, can be irreversibly transformed into an active enzyme by cleavage of covalent bonds. The proteolytic enzymes trypsin and chymotrypsin provide a classic example of zymogens and their activation. Their inactive precursor molecules, trypsinogen and chymotrypsinogen, respectively, are formed in the pancreas, where they would do damage if they were in ail active form. In the small intestine, where their digestive properties are needed, they are activated by cleavage of specific peptide bonds.

The conversion of chymotrypsinogen to chymotrypsin is catalyzed by try sin, which in turn arises from trypsinogen as a result of a cleavage reaction catalyzed by the enzyme enteropeptidase. Chymotrypsinogen consists of a single polypeptide chain 245 residues long, with five disulfide (-S-S-) bonds. When chymotrypsinogen is secreted into the small intestine, trypsin present in the digestive system cleaves the peptide bond between arginine 15 and isoleucine 16, çounting from the N-terminal end of the chymotrypsinogen sequence. The cleavage produces active π-chymotrypsin. The 15-residue fragment remains bound to the rest of the protein by a disulfide bond.

Although π-chymotrypsin is fully active, it is not the end product of this series of reactions. It acts on itself to remove two dipeptide fragments, producing a-chymotrypsin, which is also fully active. The two dipeptide fragments cleaved off are Ser 14-Arg 15 and Thr 147-Asn 148; the final form of the enzyme, α-chymotrypsiti, has three polypeptide chains held together by two of the five original, and still intact, disulfide bonds. When the term chymotrypsin is used without specifying the π or the α form, it is the final a form that is meant. The changes in primary structure that accompany the conversion of chymotrypsinogen to α-chymotrypsin bring about changes in the tertiary structure. The enzyme is active because of its tertiary structure, just as the zymogen is inactive because of its tertiary structure.

The three-dimensional structure of chymotrypsin has been determined by x-ray crystallography. The protonated amino group of the isoleucine residue exposed by the first cleavage reaction is involved in an ionic bond with the carboxylate side chain of aspartate residue 194. This ionic bond is necessary for the active conformation of the enzyme, since it is near the active site. Chymotrypsinogen lacks this bond; therefore, it does not have the active conformation and cannot bind substrate. Blood clotting also requires a series of proteolytic activations involving several proteins, particularly the conversions of prothrombin to thrombin and of fibrinogen to fibrin. Blood clotting is a complex process; for this discussion it is sufficient to know that activation of zymogens plays a crucial role. In the final, best characterized step of clot formation, the soluble protein fibrinogen is converted to the insoluble protein fibrin as a result of the cleavage of four peptide bonds. The cleavage occurs as the result of action of the proteolytic enzyme thrombin, which in turn is produced from a zymogen called prothrombin. The conversion of prothrombin to thrombin requires a number of proteins called clotting factors, as well as Ca^{2+}.

Early stages of blood clotting consist of an elaborate multistep mechanism in which the action of one clotting factor affects the behaviour of many molecules of the next factor; this cascade effect allows for fine tuning of the process but can also cause great problems if something goes wrong with one of the steps. The molecular disease hemophilia for example, is typically caused by a lack of one of the clotting factors. A hemophiliac can bleed to death from a very small cut that would not trouble a normal person.

Active sites in enzymes

Information on well-known systems such as chymotrypsin can lead to general principles that are applicable to all enzymes. Enzymes catalyse chemical reactions in many ways, but all reactions have in common the requirement that some reactive group on the enzyme interact with the substrate.

In proteins the α-carboxyl and α-amino groups of the amino acids are no longer free, since they have formed peptide bonds. Thus, the side-chain reactive groups are the ones involved in the action of the enzyme. Hydrocarbon side chains do not contain reactive groups and are not involved in the process. Functional groups that can play a catalytic role include the imidazole group of histidine, the hydroxyl group of serine, the carboxyl side chains of aspartate and glutamate, the sulfhydryl group of cysteine, the amino side chain of lysine, and the phenol group of tyrosine.

Chymotrypsin catalyzes the hydrolysis of pepticle bonds adjacent to aromatic amino acid residues in the protein being hydrolyzed; other residues are attacked at lower frequency. In addition, chymotrypsin catalyzes the hydrolysis of esters in model studies in the laboratory. The use of model systems is common in biochemistry, since a model provides the essential features of a reaction in a simple form that is easier to work with than the one found in nature.

Model systems based on the hydrolysis of esters are frequently used to study the peptide hydrolysis reaction. A typical model compound is *p*-nitrophenyl acetate, which is hydrolyzed in two stages. The acetyl group is covalently attached to the enzyme at the end of the first stage of the reaction, but the *p*-nitrophenolate ion is released. In the second stage the acyl-enzyme intermediate is hydrolyzed, releasing acetate and regenerating the free enzyme.

Serine residue at position 195 is required for the activity of chymotrypsin; in this respect chymotrypsin is typical of a class of enzymes known as serine proteases. The enzyme is completely inactivated when this serine reacts with diisopropylphosphofluoridate (DIPF), forming a covalent bond that links the serine side chain with the DIPF. The formation of covalently modified versions of specific side chains on proteins is called labeling; it is widely used in laboratory studies. The other serine residues of chymotrypsin are far less reactive and are not labeled by DIPF.

Histidine 57 is another essential amino acid residue in chymotrypsin. Chemical labeling again provides the evidence for involvement of this residue in the activity of chymotrypsin. In this case the reagent used to label the essential amino acid residue is N-tosylamido-L-phenylethyl chloromethyl ketone (TPCK), also called tosyl-L-phenylalanine chloromethyl ketone. The phenylalanine moiety is bound to the enzyme because of the specificity for aromatic amino acid residues at the active site, and the active-site histidine residue reacts because the labelling reagent is similar to the usual substrate.

Serine 195 and histidine 57 are required for the activity of chymotrypsin; therefore, they must be close to each other in the active site. The determination of the three-dimensional structure of the enzyme by x-ray crystallography provides evidence that the active-site residues do indeed have a close spatial relationship. The folding of the chymotrypsin backbone, mostly in an antiparallel pleated sheet array, positions the essential residues around an active-site pocket. Only a few residues are directly involved in the active site, but the whole molecule is necessary to provide the correct three-dimensional arrangement for those critical residues.

Other important pieces of information about the three-dimensional structure of the active site emerge when a complex is formed between chymotrypsin and a substrate analog. When one such substrate analog, formyl-L-tryptophan, is bound to the enzyme, the tryptophan side chain fits into a hydrophobic pocket near serine 195. This type of binding is not surprising in view of the specificity of the enzyme for aromatic amino acid residues at the cleavage site. The results of x-ray crystallography show, in addition to the binding site for aromatic amino acid side chains of substrate molecules, a definite arrangement of the amino acid side chains that are responsible for the catalytic activity of the enzyme.

Serine 195 and histidine 57, are involved in the mechanism of catalytic action. The oxygen of the serine side chain is a nucleophile, or nucleus-seeking substance. A nucleophile tends to bond to sites of positive charge or polarization (electron-poor sites) in contrast to an electrophile, or electron-seeking substance, which tends to bond to sites of negative charge or polarization (electron-rich sites). The nucleophilic oxygen of the serine attacks the carbonyl carbon of the peptide group. The carbon now has four single bonds, and a tetrahedral intermediate is formed; the original -C=O bond becomes a single bond, and the carbonyl oxygen becomes an oxyanion. The acyl-enzyme intermediate is formed from the tetrahedral species.

The histidine and the amino portion of the original pepticle group are involved in this part of the reaction, as the amino group hydrogen bonds to the imidazole portion of the histidine. Note that the imidazole is already protonated and that the proton came from the hydroxyl group of the serine.The histidine behaves as a base in abstracting the proton from the serine; in the terminology of the physical organic chemist, the histidine acts as a general base catalyst. The carbon-nitrogen bond of the original pepticle group breaks, leaving the acyl-enzyme intermediate.

The proton abstracted by the histidine has been donated to the leaving amino group. In donating the proton, the histidine has acted as an acid in the breakdown of the tetrahedral intermediate, although it acted as a base in its formation. In the deacylation phase of the reaction, the last two steps are reversed, with water acting as the attacking nucleophile. In this second phase the water is hydrogen bonded to the histidine. The oxygen of water now per forms the nucleophilic attack on the acyl carbon that came from the original pepticle group. Once again a tetrahedral intermediate is formed.

In the final step of the reaction the bond between the serine oxygen and the carbonyl carbon breaks, releasing the product with a carboxyl group where the original peptide group used to be and regenerating the original enzyme. Note that the serine is hydrogen bonded to the histidine. This hydrogen bond in creases the nucleophilicity of the serine, whereas in the second part of the reaction the hydrogen bond between the water and the histidine increased the nucleophilicity of the water. The mechanism of chymotrypsin action is a particularly well studied and, in many respects, a typical one. Numerous types of reaction mechanisms for enzyme action are known, and we shall discuss them in the contexts of the reactions catalyzed by the enzymes in question.

Chapter 4

Biochemistry of Nucleic Acids

The quaternary structure of nucleic acids frequently includes interaction with proteins. One well-known example is the association of RNA and proteins in ribosomes; another is the self-assembly of tobacco mosaic virus, in which the nucleic acid strand winds through a cylinder of coat protein subunits. Monomers of nucleic acids are known as nucleotides. An individual nucleotide consists of three parts—a nitrogenous base, a sugar, and a phosphoric acid residue—all of which are covalently bonded together. The order of bases in the nucleic acids specifies the genetic code.

The nucleic acid bases are of two types, pyrimidines and purines. Three pyrimidine bases, cytosine, thymine, and uracil, commonly occur. Cytosine is found in both RNA and DNA. Uracil occurs only in RNA. In DNA thymine is substituted for uracil; thymine is also found to a small extent in some forms of RNA. The common purine bases are adenine and guanine, both of which are found in both RNA and DNA.

In addition to these five commonly occurring bases, there are unusual bases with slightly different structures, that are found principally, but not exclusively, in transfer RNA. A typical nucleoside is a compound that consists of a base and a sugar covalently linked together. It differs from a nucleotide by lacking a phosphate group in its structure. In a nucleoside, a base forms a glycosidic linkage with the sugar. When the sugar is β-D-ribose, the resulting compound is a ribonucleoside; when the sugar is β-D-deoxyribose, the resulting compound is a deoxyribonucleoside.

The glycosidic linkage is from the C-1′ carbon of the sugar to the N-1 nitrogen of pyrimidines or to the N-9 nitrogen of purines. When phosphoric acid is esterified to one of the hydroxyl groups of the sugar portion of a nucleoside, a nucleotide is formed. A nucleotide is named for the parent nucleoside, with the suffix monophosphate added; the position of the phosphate ester is specified by the number of the carbon atom at the hydroxyl group to which it is esterified—for instance, adenosine 3′-monophosphate, deoxycytidine 5′-monopliosphate. The 5′ nucleotides are the ones most commonly encountered in nature.

The polymerization of nucleotides gives rise to nucleic acids. The linkage between monomers in nucleic acids involves formation of two ester bonds by phosphoric acid. The hydroxyl groups to which the phosphoric acid is esterified are those bonded to the 3′ and 5′ carbons on adjacent residues. The resulting repeated linkage is a 3′,5′-phosphodiester bond.

The nucleotide residues of nucleic acids are numbered from the 5′ end, which normally carries a phosphate group, to the 3′ end, which normally has a free hydroxyl group.

The sugar-phosphate backbone repeats itself down the length of the chain. The most important feature of the structure of nucleic acids is the identities of the bases. The sequence of bases in DNA carries the information for the genetic code. The blueprint for every living organism is contained in its DNA, and with DNA sequences becoming available in greater numbers, it is more apparent than ever that all living things are related.

Structure of DNA

The double helix structure of DNA was proposed by Watson and Crick in 1953, it touched off a flood of research activity, leading to great advances in molecular biology. The determination of the double-helical structure was based on chemical analysis of DNA base composition and on x-ray diffraction patterns. Both of these lines of evidence were necessary for the conclusion that DNA consists of two polynucleotide chains wrapped around each other to form a helix.

Hydrogen bonds between bases on opposite chains determine the alignment of the helix, with the paired bases lying in planes perpendicular to the helix axis. The sugar-phosphate backbone is the outer part of the helix. The chains run in antiparallel directions, one 3′ to 5′ and one 5′ to 3′. The helical structure of DNA can demonstrate the x-ray diffraction pattern. The combination of evidence from x-ray diffraction and chemical analysis led to the conclusion that the base pairing is complementary, meaning that adenine pairs with thymine, and guanine with cytosine. By 1953, studies of the base composition of DNA from many species had already shown that, to within experimental error, the mole percentages of adenine and thymine were equal; the same was found to be the case with guanine and cytosine. An adenine-thymine (A-T) base pair has two hydrogen bonds between the bases; a guanine-cytosine (G-C) base pair has three. Inside diameter of the sugar-phosphate backbone of the double helix is about 11 Å (1.1nm).

The distance between the points of attachment of the bases to the two strands of the sugar-phosphate backbone is the same for the two base pairs (A-T and G-C), about 11 Å (1.1 nm), so these base pairs are the right size to fit into the helix. Base pairs other than A-T and G-C are possible, but they do not have the correct hydrogen bonding pattern (e.g., A-C or G-T pairs) or the right dimensions (purine-purine or pyrimidine-pyrimidine pairs) to fit the inside diameter of the double helix.

The outside diameter of the helix is 20 Å (2 nm). The length of one complete turn of the helix along its axis is 34 Å (3.4 nm), containing ten base pairs. The atoms that make up the two polynucleotide chains of the double helix do not completely fill an imaginary cylinder around the double helix, but they leave empty spaces known as grooves. There is a large major groove and a smaller minor groove in the double helix; both can be sites at which drugs or polypeptides bind to DNA. At neutral, physiological pH, each phosphate group of the backbone carries a negative charge. Positively charged ions such as Na^+ or Mg^{2+} and polypeptides with positively charged side chains are frequently associated with DNA as a result of electrostatic attraction; eukaryotic DNA, for example, is complexed with histones, which are positively charged proteins, in the cell nucleus.

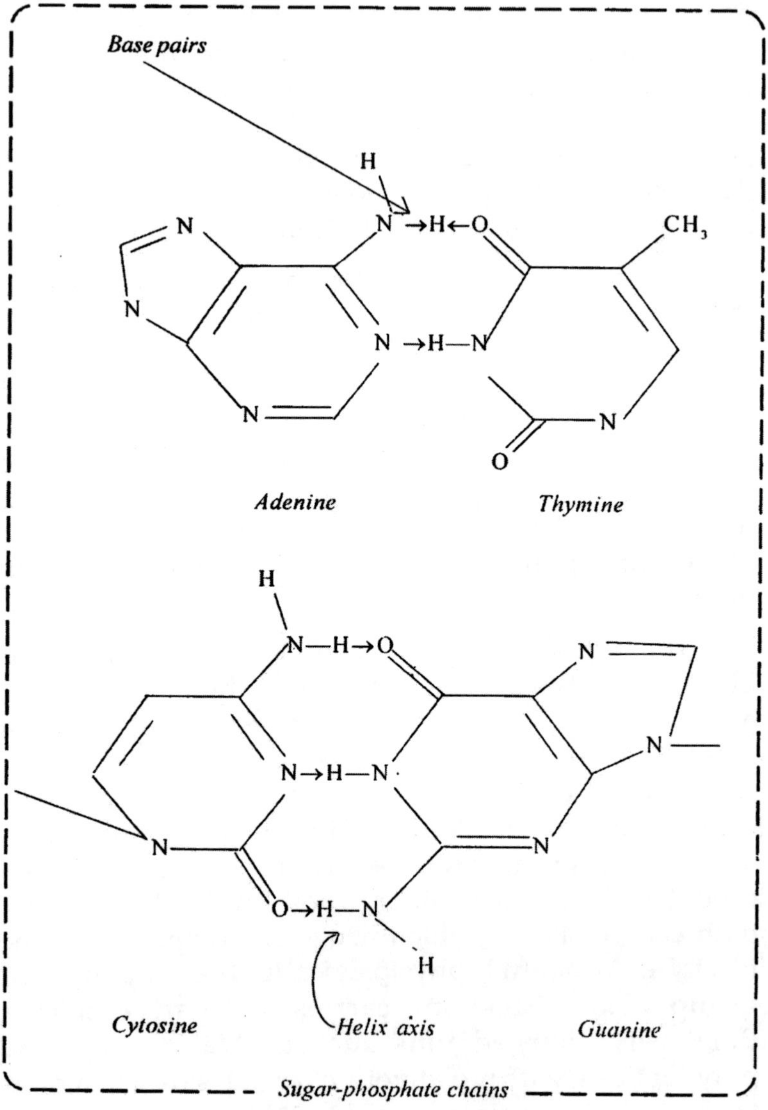

Fig. 1. Strvcture of DNA.

The antiparallel direction of the two polynucleotide chains is an aspect of the complementarity of the two strands. The form of DNA that we have been discussing so far is called B-DNA. It is thought to be the principal form that occurs in nature. However, other secondary structures can occur, depending on conditions such as the nature of the positive ion associated with the DNA. One of those other forms is A-DNA, which has 11 base pairs for each turn of the helix. Its base pairs are not perpendicular to the helix axis but lie at an angle of about 20 degrees to the perpendicular, like the blades of a propeller. An important shared feature of A-DNA and B-DNA is that both are right-handed helices; that is, the helix winds upward in the direction in which the fingers of the right hand curl when the thumb is pointing upward.

A variant form of the double helix, Z-DNA, is left-handed; it winds in the direction of the fingers of the left hand. Z-DNA is known to occur in nature, most often when there are repeated G-C sequences, but its function has not been determined at this writing. It may play a role in the regulation of gene expression. The Z form of DNA is also a subject of active research among biochemists who want to design drugs that will bind specifically to it.

Supercoiling

Twisting coiling or supercoiling of the double helix is possible. The first example of supercoiling we shall consider is the case in which DNA is not complexed to proteins. This naked DNA occurs in prokaryotes. If the sugar-phosphate backbone of a prokaryotic DNA forms a covalently bonded circle, the structure is still relaxed. Some extra twists are added if the DNA is unwound slightly before the ends are joined to form the circle. A strain is introduced in the molecular structure, and the DNA assumes a new conformation to compensate for the unwinding. If, because of unwinding, a right-handed double helix acquires an extra left-handed helical twist, the circular DNA is said to be negatively supercoiled.

Under different conditions it is possible to form a right-handed, or positively supercoiled structure in which there is overwinding of the closed-circle double helix. The difference between the positively and negatively supercoiled forms lies in their right and left-handed natures, which in turn depend on the overwinding or underwinding of the double helix. Enzymes that affect the supercoiling of DNA have been isolated from a variety of organisms.

Naturally occurring circular DNA is negatively supercoiled except during replication, when it becomes positively supercoiled. It is critical for the cell to have control over this process. Two classes of enzymes are involved in the regulation of the process, one to relax the supercoil and one to rewind it. Topoisomerases are enzymes that hydrolyze a phosphodiester linkage in one strand of the double helix, relax the supercoiling by rotating one strand around the other, and then reseal the break.

DNA gyrases induce negative supercoiling in relaxed, closed-circular DNA. These enzymes play an important role in the replication of DNA. Supercoiling has been observed experimentally in naturally occurring DNA. Particularly strong evidence has come from electron micrographs that clearly show coiled structures in circular DNA from a number of different sources, including bacteria, viruses, mitochondria, and chloroplasts. Supercoiled DNA can be detected experimentally by ultracentrifugation, since it sediments more rapidly than the relaxed form. It has been known for some time that prokaryotic DNA is normally circular, but supercoiling is a relatively recent subject of research.

Computer modeling has helped scientists visualize many aspects of the twisting and knotting of supercoiled DNA by obtaining stop-action images of very fast changes. The supercoiling of the nuclear DNA of eukaryotes is more complicated than the supercoiling of the circular DNA from prokaryotes. Eukaryotic DNA is complexed with a number of proteins, especially with basic proteins that have abundant positively charged side chains at physiological pH.

Electrostatic attraction between the negatively charged phosphate groups on the DNA and the positively charged groups on the proteins favours the formation of complexes of this sort. The resulting material is called chromatin. The supercoiling of chromatin must accommodate the presence of the proteins. The principal proteins in chromatin are the histones, of which there are five main types, called H_1, H_2A, H_2B, H_3, and H_4. All these proteins contain large numbers of basic amino acid residues, such as lysine and arginine. In the chromatin structure, the DNA is tightly bound to all the types of histone but H_1.

The H_1 protein is comparatively easy to remove from chromatin, but dissociating the other histories from the complex is more difficult. Proteins other than histories are also complexed with the DNA of eukaryotes, but they are neither as abundant nor as well studied as histones. Chromatin resembles beads on a string, in electron micrographs. This appearance reflects the molecular composition of the protein-DNA complex. Each 'bead' is a nucleosome, consisting of DNA wrapped around a histone core. This protein core is an octamer, which includes two molecules of each type of histone but H_1; the composition of tile octamer is $(H_2A)_2(H_2B)_2(H_3)_2(H_4)_2$.

The 'string' portions are called spacer regions; they consist of DNA complexed to some HI histone and nonhistone proteins. As the DNA coils around the histories in the nucleosome, about 150 base pairs are in contact with the proteins; the spacer region is about 30 to 50 base pairs long.

Denaturation of DNA

The hydrogen bonds between base pairs are an important factor in holding the double helix together. The amount of stabilizing energy associated with the hydrogen bonds is not great, but the hydrogen bonds hold the two polynucleotide chains in the proper alignment. In addition, the stacking of the bases in the native conformation of DNA contributes a large part of the stabilization energy. Energy must be added to a sample of DNA to break the hydrogen bonds and disrupt the stacking interactions. This is usually carried out by heating the DNA in solution.

The heat denaturation of DNA, also called melting, can be monitored experimentally by observing the absorption of ultraviolet light. The bases absorb light in the 260-nm wavelength region.

As the DNA is heated and the strands separated, the wavelength of absorption does not change, but the amount of light absorbed increases. This effect is called hyperchromicity. It is based on the fact that the bases, which are stacked on top of one another in native DNA, become unstacked as the DNA is denatured. Since the bases interact differently in the stacked and unstacked orientations, their absorbance changes. Heat denaturation is a way to obtain singlestranded DNA, which has many uses.

One particularly important use of single-stranded DNA is as the starting point for sequence determination, the most ambitious example of which is described in the box. Under a given set of conditions, there is a characteristic midpoint of the melting curve for DNA from each distinct source. The underlying reason for this property is that each type of DNA has a given well-defined base composition. A G-C base pair has three hydrogen bonds, and an A-T base pair has only two.

The higher the percentage of G-C base pairs, the higher the melting temperature of a DNA molecule. Renaturation of denatured DNA is possible with slow cooling. The annealed, separated strands can recombine and form the base pairs responsible for maintaining the double helix.

RNAs and their structures

There are three kinds of RNA, transfer RNA (tRNA), ribosomal RNA (rRNA), and messenger RNA (mRNA), play an important role in the life processes of cells. All three participate in the synthesis of proteins in a series of reactions ultimately directed by the base sequence of the cell's DNA. The base sequences of all types of RNA are determined by !hat of DNA. The process by which the order of bases is passed from DNA to RNA is called transcription. Ribosomes, in which rRNA is associated with proteins, are the sites for assembly of the growing polypeptide chain in protein synthesis.

Amino acids are brought to the assembly site covalently bonded to tRNA, as aminoacyl tRNAs. The order of bases in mRNA specifies the order of amino acids in the growing protein; this process is called translation of the genetic message. A sequence of three bases in mRNA directs the incorporation of a particular amino acid into the growing protein chain.

Transfer RNA

Transfer RNA (tRNA) is the smallest of the three important kinds of RNA. Dozens of different species of tRNA molecules can be found in every living cell, since there is at least one tRNA that bonds specifically to each of the amino acids that commonly occur in proteins. Frequently there are several tRNA molecules for each amino acid. A tRNA is a single-stranded polynucleotide chain, usually about 80 nucleotide residues long, and generally having a molecular weight of about 25,000 daltons. Intrachain hydrogen bonding occurs in tRNA, forming A-U and G-C base pairs similar to those that occur in DNA except for the substitution of uracil for thymine.

The duplexes thus formed have the A-helical form, rather than the B-helix, which is the predominant form in DNA. The molecule folds back on itself in a kind of cloverleaf structure, which can be considered the secondary structure of tRNA. The hydrogen-bonded portions of the molecule are called stems, and the non-hydrogen-bonded portions are loops. Some of these loops contain modified bases. During protein synthesis, both tRNA and mRNA are bound to the ribosome in a definite spatial arrangement that ultimately ensures the correct order of the amino acids in the growing polypeptide chain.

A particular tertiary structure is necessary for tRNA to interact with the enzyme that covalently attaches the amino acid to the 3′ end. To produce this tertiary structure, the cloverleaf folds into an L-shaped conformation that has been determined by x-ray diffraction.

Ribosomal RNA

rRNA molecules tend to be quite large, and only a few types of rRNA exist in a cell. Because of the intimate association between rRNA and proteins, a useful approach to understanding the structure of rRNA is to investigate ribosomes themselves. The RNA portion of a ribosome accounts for 60% to 65% of the total weight, and the protein portion constitutes the remaining 35% to 40% of the weight.

Dissociation of ribosomes into their components has proved to be a useful way of studying their structure and properties. A particularly important endeavour has been to determine both the number and the kind of RNA and protein molecules that make up ribosomes. This approach has helped elucidate the role of ribosomes in protein synthesis. In both prokaryotes and eukaryotes, a ribosome consists of two subunits, one larger than the other. In turn, the smaller subunit consists of the large RNA molecule and about 20 different proteins; the larger subunit consists of 2 RNA molecules in prokaryotes (3 in eukaryotes) and about 35 different proteins in prokaryotes (about 50 in eukaryotes).

The subunits are easily dissociated from one another in the laboratory by lowering the Mg^{2+} concentration of the medium. Raising the Mg^{2+} concentration to its original level reverses the process, and active ribosomes can be reconstituted by this method. A technique called analytical ultracentrifugation has proved very useful for monitoring the dissociation and reassociation of ribosomes. We need not consider all the details of the method as long as it is clear that the basic principle of the experiment is observation of the motion of ribosomes, RNA, or protein in a centrifuge. Both the size and the shape of a particle determine how fast it will move toward the bottom of the tube.

The motion of the particle is characterized by a sedimentation coefficient, expressed in Svedberg units (S), which are named after Theodor Svedberg, the Swedish scientist who invented the ultracentrifuge. The S value increases with the molecular weight of the sedimenting particle, but it is not directly proportional to it, because the particle's shape also affects its sedimentation rate. Ribosomal RNA have been studied extensively using sedimentation coefficients. Most research on prokaryotic systems has been done with the bacterium *Escherichia coli,* which we shall use as an example here.

An *E.coli* ribosome typically has a sedimentation coefficient of 70S. When an intact 70S bacterial ribosome dissociates, it produces a light 30S subunit and a heavy 50S subunit. Note that the values of sedimentation coefficients are not additive, showing the dependence of the S value on the shape of the particle. The 30S subunit contains a 16S rRNA and 21 different proteins. The 50S subunit contains a 5S rRNA, a 23S rRNA, and 34 different proteins. For comparison, eukaryotic ribosomes have a sedimentation coefficient of 80S, and the small and large subunits are 40S and 60S, respectively.

The small subunit of eukaryotes contains an 18S rRNA, and the large subunit contains three types of rRNA molecules: 5S, 5.8S, and 28S. The 5S rRNA has been isolated from many different types of bacteria, and the nucleotide sequences have been determined. A t-,pical 5S rRNA is about 120 nucleotide residues long and has a molecular weight of about 40,000 daltons. Some sequences have also been determined for the 16S and 23S rRNA molecules. These larger molecules are about 1500 and 2500 nucleotide residues long, respectively.

The molecular weight of 16S rRNA is about 500,000 daltons, and that of 23S rRNA is about 1 million daltons. There appear to be considerable degrees of secondary and tertiary structure in the larger RNA molecules. A secondary structure has been proposed for 16S rRNA, and suggestions have been made about the way in which the proteins associate with the RNA to form the 30S subunit. The self-assembly of ribosomes takes place in the living cell, and the process can be duplicated in the laboratory. Elucidation of ribosomal structure is an active field of research.

Chapter 5

Techniques of DNA Looping

DNA looping was an unanticipated discovery. DNA looping permits additional proteins to bind in the vicinity of the complex and to interact with it. Multiple proteins could also assist or interfere with the interactions necessary for looping. Thus, a rich diversity of regulation systems can be expected to utilize DNA looping. Multiple proteins can be required to regulate properly the activity of a complex on DNA.

For example, transcription of a gene could require the simultaneous presence of four different conditions, the status of each of which must be transmitted by a protein sensor to the initiation complex. In order that the protein sensors, which we normally call regulatory proteins, confine their activities to the correct genes, they bind to specific sequences located near the genes to be regulated. There is space for only two or three proteins to bind to DNA alongside an initiation complex. DNA looping also generates cooperativity in the binding of a looping protein to the two DNA sites involved. When a protein has bound to one of the two DNA sites, it is held near the other DNA site because the furthest the two sites can separate is the length of the DNA between the two.

This tethering of one site to another increases the concentration of the protein near the second site and generates cooperativity in the binding to the two sites—whenever one site is bound by the protein, the other tends also to be bound. The cooperativity resulting from DNA looping permits proteins to saturate their binding sites even though the protein concentration is well below the dissociation constant of the protein from an individual site. Thus, the microscopic dissociation of the protein from a single site can still occur, but overall, the protein remains at the two sites involved.

Another way to view this phenomenon is that the protein does riot come free of the DNA and diffuse away from the vicinity of the binding sites unless it lets go of both DNA-binding sites simultaneously. As a result, comparatively low concentrations of proteins are required to saturate DNA-binding sites involved with looping, and the sites need not be constructed to possess particularly high affinities for the proteins. Why would nature want to saturate DNA sites while using low concentrations of protein, and without building the DNA sites with particularly high affinities? One answer seems to be that a large number of regulatory proteins, perhaps 50,000 in a eukaryotic cell, must simultaneously be present in the nucleus.

Therefore, the concentration of any single one must be low, perhaps 10^{-10} M. While this concentration of protein could saturate sites with dissociation constants of 10^{-11} M, binding many proteins to DNA with such a high affinity could well interfere with other DNA processes such as replication and recombination. DNA looping takes care of this problem through the cooperativity. The DNA sites involved can have individual dissociation constants for the protein much lower, say 10^{-9} M, but by virtue of looping, pairs of such sites would still be nearly fully occupied.

Thus, by virtue of looping, the objectives of low concentrations of the proteins, binding sites that do not weld the proteins to the DNA, and nearly complete occupancy of the binding sites can all be simultaneously achieved. At one extreme DNA looping is generated by one protein with two DNA-binding sites. At the other extreme, two different proteins bind the two different sites, and because they are then held near one another by the DNA, they bind to each other and form a loop. Therefore, we can expect DNA looping to be generated by a single protein, by an oligomeric protein, or by two proteins that only associate at high concentrations or in the presence of DNA containing their DNA-binding sites.

Brief history and overview

In fact, it was sufficiently unexpected that a number of clues to its existence were overlooked in the years preceding its discovery. When it was finally proposed and presented with strong supporting evidence, the idea was sufficiently alien that it was not readily accepted for publication. Once reasonably solid data did get published, however, acceptance of looping came rapidly because it aided understanding of several widely observed phenomena.

The clue that motivated the experiments that ultimately uncovered DNA looping was an observation made by Englesberg on the arabinose operon in *Escherichia coli*. He obtained genetic evidence for a site lying upstream of the *araBAD* promoter that acted to depress activity of the promoter. A long series of studies followed that were aimed at determining the mechanism by which repression in the ara system could operate from the upstream site. Ultimately, it was found that this was accomplished by protein bound to the upstream site as well as to a site beside the promoter 210 base pairs downstream. Subsequently, DNA looping has been shown to function in a wide variety of other systems, both prokaryotic and eukaryotic. The objective of this review is to give the reader an understanding of how and why looping occurs, and what it has been found to accomplish. First, the various assays that have been used to detect and measure looping are described. These different methods give an idea of the richness and diversity of the looping phenomena, as they range from detecting changes in the conformation of the DNA involved in looping to measuring changes in the regulation of genes in growing cells. As the heart of looping is the physical chemistry, an overview of the physical basis for looping is given. Then, a number of important biological systems are described in which looping plays a prominent role. This review does not contain an exhaustive list of references relevant to DNA looping; the available electronic bibliographic databases can provide this for the interested reader.

Assays of Looping

A variety of assays have been developed to detect and quantitate DNA looping. The first of these, used with the intention of demonstrating looping, was the helical-twist experiment in the arabinose operon. In this type of experiment the two sites involved in looping are rotated around the axis of the cylindrical DNA with respect to one another by the insertion or deletion of differing amounts of DNA between the sites.

If the two sites are positioned for easiest looping, insertion of five base pairs between them rotates one site with respect to the other half a turn around the DNA. As a result, the formation of a loop now requires not only bending the DNA to bring the sites near one another, but also twisting of the DNA to bring the sites to the correct face of the DNA. Because DNA resists torsional stress, such a positioning increases the energetic costs of looping, thereby decreasing its frequency of occurrence. Of course, the amount of interference with looping depends upon the size, shape, and flexibility of the proteins involved, the distance separating the binding sites, and the energies available for looping.

The most convincing helical-twist experiments measure the responses to large numbers of spacing changes and show multiple cycles of oscillations as the sites are positioned alternatingly to favour and disfavour looping. The conceptual basis for helical-twist experiments comes from DNA cyclization experiments first done by Shore, Langowski, & Baldwin. For DNA several hundred base pairs in length, the cyclization rate of linear molecules to covalently closed circles in the presence of DNA ligase depended not only on the total length of the DNA, but also on the relative phases of the short single-stranded ends. Misphasing the ends by half a helical turn dramatically decreased the cyclization rate.

Electrophoresis of looped small circles

The binding of proteins to sites located on small supercoiled DNA molecules of about 400 base pairs, as well as electrophoresis of the complexes, has proven to be a particularly useful tool for studying DNA looping. The power of the assay results from the fact that weakly looping systems can be studied. This results from the fact that the supercoiling places the sites in very close proximity even in the absence of the protein. As long as the sites are located on the proper face of the DNA, little of the protein-DNA or protein- protein binding energy needs to be expended to bend the DNA or position the sites so that the protein can simultaneously contact both. The assay built upon gel retardation experiments in which the binding of a protein to DNA can be detected by performing electrophoresis under conditions where the protein remains bound during the electrophoretic run. Normally in these experiments the protein-DNA complex migrates more slowly than the DNA molecules free of protein. On small supercoiled circles such retardation can be seen when the DNA contains only a single binding site for a protein. When looping between two sites on the circle occurs, however, the complex may migrate faster than either the free DNA or the DNA with protein bound to a single site. The reason for the mobility increase when two operators are present must be that looping between the two sites holds the DNA in a highly compact structure.

Although the dependence on both protein-binding sites for the existence of the altered mobility species suggests DNA looping, additional properties strengthen this conclusion. First, the complex is much more stable when two binding sites are present than when only a single site is present. That is, its dissociation rate is much reduced when the two sites are present. Second, the increased stability and faster mobility properties show helicaltwist effects when one site is rotated to the opposite face of the DNA.

Generation of sandwich (crossjoined) structures

In DNA looping, a protein or complex of proteins binds to two DNA sites. There is no fundamental reason that the same two sites could not be on two different DNA molecules. Then, the DNA molecules would be held together by the protein(s), and a sandwich structure would be formed. The detection of such sandwiches is easy in the gel retardation assay in which proteins remain bound to DNA during the electrophoresis. Unfortunately, such sandwiches are hard to form. As described above, one reason looping occurs is because the two DNA sites are held in the vicinity of one another by the intervening DNA. This greatly reduces the entropic costs of loop formation. When the two sites are on different molecules, the entropic costs of forming the sandwich are high. Therefore many looping systems lack sufficient binding energy to form sandwiches, although their presence has occasionally been used to support conclusions that a system loops.

Structure of DNA in loops

Looped DNA has to be bent. When the loop is small enough, the bending will be great enough to affect substantially the widths of the major and minor grooves of the DNA. These width changes can be revealed by DNAsel digestion since the activity of this enzyme is sensitive to the width of the minor groove.

The regions facing inward will be so compressed that the enzyme will have little activity there, but portions of the regions facing outward will be suitably spaced for rapid attack by the enzyme, and will be readily cleaved. In some cases then, the pattern of sites that have gained and lost DNAsel sensitivity reflects the helical repeat of the DNA and shows periodicities of 10 or 11 base pairs.

A dramatic example of this phenomenon is an artificial looping situation involving lambda phage repressor. Normally the repressor dimers bind to DNA alongside one another and also interact with one another via protein-protein interactions. These protein-protein interactions are sufficiently strong that if the binding sites are moved 55 base pairs apart, the proteins will still interact with one another and in doing so, will form a DNA loop. The region between the binding sites shows areas of strongly protected and enhanced DNAse sensitivity. Looping can alter not only the conformation of DNA lying between the two protein-binding sites, but it can also alter DNA beyond the looped region.

The holding of the two protein-binding sites near

one another directs the segments of DNA coming into the loop and leaving the loop to lie in similar directions. This constraint tends to hold the ends of the DNA near one another. Such a proximity increase can increase the rate at which the ends will ligate together when DNA ligase is added. Such ligation rate effects have been observed in a looping system comprising a DNA replication protein from plasmid RP6.

Electron microscopic observation of looping

DNA looping formed by protein binding to two sites separated by more than about 50 base pairs should be readily observable by electron microscopy. Such looping has been observed in a number of systems. Although seeing is close to believing, the labour of obtaining statistically valid data has greatly limited the use of microscopy in looping studies. Careful control experiments are also necessary in this type of experiment because random aggregation of protein that has bound to the correct sites can generate a loop, but one without biological significance.

Cooperativity from a distance

As described earlier, the binding of a bivalent protein to one site on the DNA increases the concentration of the protein's second DNA-binding site to nearby DNA sequences. Simultaneous binding to a second site generates looping. The affinity of the second site for the protein can be so weak and the concentration of the protein so low that the site is occupied only by virtue of looping. Another way to describe the phenomenon is that binding to the two DNA sites shows cooperativity.

The presence of one binding site increases the binding to a second site. Chemists call this phenomenon the chelate effect, and its origins can equally well be discussed in terms of the decreased entropy. The existence of cooperativity has been used as an argument for DNA looping in a number of systems. The cooperativity can be observed physiologically, as for example, in the lac operon where a mutant lac operator binds repressor so weakly that only minimal repression is observed. The introduction of a second operator in the vicinity increases the repressor occupancy of the first operator.

Tethering

DNA looping is an attractive means by which a protein bound at one site can interact with a DNA site or protein bound to a DNA site located within a few hundred or few thousand base pairs. Instead of looping, some have suggested that two nearby sites might interact by direct communication along the DNA separating the sites through crawling, threading, or conformational telegraphic mechanisms.

The two basic options of looping vs DNA-mediated communication can be distinguished by tethering two sites near one another without providing a direct DNA connection between the two. In this case, communication between the sites should be blocked if normally they communicate by mechanisms directly utilizing the intervening DNA, but communication between the sites will not be blocked if they normally utilize DNA-looping mechanisms. Two ways of fastening sites near one another have been used. These are (a) use of interlocked DNA circles, called catenanes, and (b) linking DNA by a protein bridge. Interlocked circles have been used in vitro with the *gln* system to demonstrate that stimulation of' transcription by NtrC was via looping.

The NtrC-binding site was placed on one circle and the promoter on the other. The promoter was activated by NtrC despite being bound to a different DNA molecule than the promoter. Analogous experiments were performed with the enhancer and promoter for ribosomal RNA synthesis in *Xenopus*, although there the interlocked circles were injected into oocytes for assay of enhancement. In these experiments the topological interlocking was removed in vivo within 30 minutes, but the enhancement persisted for more than 18 hours. This implies that a stable initiation stimulating complex was set up early on. DNA molecules can also be linked by a protein bridge of avidin or streptavidin binding to biotinylated ends of DNA molecules. In this case too, the enhancer worked fine.

Physical chemistry of DNA looping

Too many effects complicate in vivo DNA-looping situations for us to be able to perform meaningful calculations on looping probabilities in growing cells. For some cases, however, we can estimate whether looping will be more or less favoured by certain changes in the system. In looping, the protein-binding sites must approach within a distance equal to the size of the looping protein and the orientation of the DNA at the sites must be constrained as well.

Therefore DNA looping is closely related to ring closure of DNA molecules, and a number of results from physical chemistry on ring closure are applicable to or at least provide insight into this aspect of looping particular looping in vitro. Below are listed a few of the factors that clearly affect looping. We also discuss a few of the calculations related to looping with the objectives of seeing the consequences of changes in DNA length, lateral stiffness, and torsional stiffness.

Length and lateral stiffness of the DNA Two sites can only interact if the DNA is sufficiently flexible and the sites are sufficiently far apart that the DNA can bend to position the two sites the proper distance apart in three dimensions. The stiffer the DNA, the more difficult it is to form small loops. A measure of the stiffness Of Polymers is their persistence length. Roughly this can be thought of, within a factor of two, as the lengths of inflexible rods connected by flexible joints that approximate the statistical behaviour of the Polymer.

Therefore, bending the DNA into a circle of size near the persistence length would be very difficult because of the stiffness of the DNA. The formal definition of persistence length comes from a consideration of the statistics of polymers. Consider a collection of polymers with one end at the origin. The average of the locations of the other end will also lie at the origin. Now consider orienting each of the polymers so that the initial direction of the polymer is along the x axis. Then, on average, the other ends of the polymers will be displaced some distance in the x direction.

The amount of this displacement as the polymer length approaches infinity is defined as the persistence length. Torsional stiffness of the DNA If two sites are misoriented around the DNA, twisting of the DNA between the sites may be required to orient the sites properly for looping. On the other band, if the sites are already properly oriented, torsional stiffness of the DNA holds the sites in the correct orientation and increases the ease of loop formation.

Flexibility of the looping proteins If the looping protein is inflexible, then the DNA must assume exactly the correct conformation for looping to occur. A highly flexible protein would tolerate a large collection of DNA conformations and would loop more easily. Such a protein would not require the DNA to bend in the limit of small loops, ones in which the protein can span the distance between the two DNA sites.

The global structure of supercoiled DNA Points well separated from One another On a Circle of 5000 base pairs are brought close to one another by supercoiling of the circle. In general the effective concentration of one point near another is increased by supercoiling. This concentration increase facilitates loop formation, but the magnitude of the increase decreases as the separation between sites becomes very large.

Additional second-order effects Proteins not involved with a particular looping reaction could affect the reaction. For example, a competing looped structure could hold the DNA such that the sites required for the first looping reaction could not occur. Alternatively, a protein could bind within a region that normally would be included in a loop and prevent loop formation if the length of the DNA is insufficient to reach around the interloper. DNA kinks or bends specified by sequence could help or hinder formation of small loops by altering the amount of bend required to loop. They could also affect formation of larger loops by altering the conformations available to the DNA and therefore alter the looping probability.

Including lateral stiffness

In the limit of small circles where the size of the circle approaches the persistence length of the DNA, circle formation becomes energetically more and more expensive as the circle size decreases. The persistence length of double-stranded DNA in physiological buffers is about 450 Å. In the other extreme, as the DNA becomes longer and longer, the concentration of on end in the presence of the other falls off, but rather slowly, and in this limit can be calculated. Due to the random-walk nature of the path of long DNA, this concentration falls as $(3/2\pi nl^2)^{-3/2}$ where l is the step length in the random walk, or interbase distance, and n is the number of steps, or bases. In the above calculation, only the presence of the two ends in the same volume element is calculated, independent of the directions from which the two ends approach one another. For some ring-closure problems and some looping problems the DNA segments should be oriented properly.

A more complicated calculation by Yamakawa & Stockmayer has considered this case. In such a situation we expect that the concentration of one end in the presence of the other will be low for short DNA, for which the necessary bending is difficult, will rise to a maximum for some intermediate length of DNA, and will then fall as $(\text{length})^{-3/2}$ as in the random-walk situation, due to the reducing probability that the two ends will occupy the same volume. The quantitative calculations bear this out.

Including torsional stiffness

In the example just considered, the torsional stiffness in the connecting string can increase or decrease the effective concentration of one site in the vicinity of the other. If the sites were properly oriented, then each collision would be a productive collision and functionally the effective concentration would be increased. The converse is true if the sites are misoriented.

The torsional stiffness of DNA is such that it costs about 2 kcal/mole to twist a stretch of 200 base pairs half a turn. Of course, the precise values depend upon the nucleotide sequence and the buffer conditions. This energetic cost is inversely proportional to the length of the DNA separating the sites. Since the typical energies involved in protein-DNA binding are 10 to 15 kcal/mole, torsional stiffness can be expected to play an important role for looping involving sites closer together than about 500 base pairs. As mentioned earlier, if the binding energies are substantially greater or if the protein is unusually flexible or extended, such helical-twist effects can be greatly reduced.

Including both lateral and torsional stiffness

The consequences to ring closure of both lateral and torsional stiffness in the DNA have been considered by Shimada & Yamakawa. Their calculation shows that the probability of ring closure for DNA of less than about 500 base pairs long oscillates with period of 10.5 base pairs while following roughly the same shape of ring-closure curve as is followed when twisting is ignored. This is the behaviour expected. When the ends are in phase with one another. the rate of ring closure is high, and when they are out of phase with each other, twisting of the DNA is required in addition to bringing the ends together, and the rate is reduced.

Monte carlo calculations

The ring-closure problem for a laterally and torsionally stiff polymer is just about as complex a problem as can be solved analytically. For some of the biological problems, however, we will need to consider the effects of natural bends in the DNA, the consequences of other proteins bound to the DNA, and of supercoiling of the DNA. In these cases, Monte Carlo calculations are the best computational approach. In such approaches one computes over and over, millions of times, the behaviour of a system when particular randomly chosen values are applied to each of the variables that describe the system.

For example, in a Monte Carlo calculation of DNA conformation, the variables would be the various angles describing the path of the phosphodiester backbone, the angles describing the sugar conformation, and the angle of the base-sugar bond. The values assigned to the random variables are weighted in accordance with the distribution that occurs in nature. For example, if there is complete freedom of rotation about a particular bond, then any value between 0 and 360 degrees will be randomly chosen for the angle. On the other hand, if the bond is constrained to lie near 10 degrees, most of the time the angles will be chosen to lie near 10 degrees. The probability of deviating an amount δ from 10 degrees would be appropriately weighted. In performing a Monte Carlo calculation of the conformation of DNA, the path of a DNA molecule is then computed for a set of the randomly chosen angles. This is done over and over, and the fraction of conformations that forms loops is obtained.

This can be converted to an effective concentration of one end in the presence of the other. Levene & Crothers did this ignoring the torsion angle, and later Hagerman included computation of the torsion angle. When applied to situations similar to those for which the analytical solutions of Shimada & Yamakawa apply, the results closely agree, suggesting that both the analytical and numerical approaches are correct (or most unlikely, both are incorrect).

Looping in gene regulation

This section describes and surveys a number of the most carefully studied DNA looping systems. Both an historical and a mechanistic perspective is taken in these descriptions. The systems illustrate the diversity of the uses to which nature has put DNA looping and show the generality of looping.

Prokaryotic systems

In the absence of arabinose, the uninduced level of the *araBAD* genes was about 10 times normal. That is, the deletion appeared to remove a site through which AraC protein acted to repress its own activation of the ara_{PBAD} promoter. Presumably, in the absence of arabinose, a small fraction of AraC would be in the inducing state, and would weakly activate transcription of the $arap_{BAD}$ promoter if the site required for repression by AraC were absent. Englesberg's deletion would remove this site. At about the time the *ara* operon was being probed genetically, biochemical studies were unraveling the *lac* operon. There repression occurred through steric hindrance in the binding of repressor or RNA polymerase.

Therefore the possibility that a repressor could function from a position upstream of all the sites that are required for induction, as the genetic data in the ara system indicated, was not easily understandable. This led to an extensive study of the phenomenon of repression from upstream. Standard genetics reproduced the upstream repression phenomenon. Subsequently, locating the AraC protein-binding sites on the DNA led to the mistaken conclusion that repression was mediated by the *araO*$_1$ site. This is located immediately adjacent to the initiation complex. A protein bound in such a position can be imagined to be capable of generating repression by a mechanism involving direct side-by-side touching of proteins all in a row along the DNA; from AraC protein bound at *araO*$_1$ contacting the cyclic AMP receptor protein contacting AraC protein bound at araI finally contacting RNA polymerase bound at the promoter.

Genetic engineering techniques eventually permitted more accurate mapping of the upstream site required for repression of the arabinose operon. It was found not to be *araO*$_1$, but another site that lies more than 200 base pairs upstream from the induction complex at the promoter, called *araO*$_2$. That is, a deletion ending as far as 200 base pairs upstream from P_{BAD} eliminated the self-repression by *AraC* protein. Clearly some signal was being sent from *araO*$_2$ to the initiation complex at $P_{BAD,}$ and the domino mechanism of direct protein-to-protein interaction for transmitting a signal from one site to another by a continuous row of proteins on the DNA was impossible. Amongst the possibilities for transmission of the signal was DNA looping whereby *Ara*C protein bound at the *ara*O_2 site would contact *Ara*C protein bound at the araI site at P_{BAD}. Helical-twist experiments were used to demonstrate that indeed, looping was the mechanism of interaction between *ara*O_2 and *araI*. Inserting five base pairs at an irrelevant site between *ara*O_2 and araI effectively rotates these sites one half a turn of the DNA helix with respect to one another. If previously the sites faced one another and looping required bringing proteins bound to these sites into contact, the insertion of five bases required not only bringing the proteins together, but also twisting the DNA half a helical turn.

The energetic costs of such a twist are about 2 kcal/mole, which in the arabinose operon were sufficient to interfere substantially with looping. Inserting 10 base pairs at the same site restored looping. The only reasonable mechanism compatible with these observations is DNA looping. One virtue of the arabinose system was the relative simplicity of utilizing genetics to study looping. One might expect that mutations in both the DNA and in the AraC protein would interfere with looping. Both types have been found and studied. Repression-negative mutations in the DNA were confined to the two AraC protein-binding sites, araO_2 and araI. Those in araO_2 reduced the binding affinity to this site about 30-fold. Those in araI were more complicated, and originally appeared to decrease the ability of AraC bound at araI to repress without affecting the binding ability of AraC.

More accurate binding measurements have shown they slightly increase the binding affinity of AraC protein and they may be constitutive in nature. Mutations in AraC have also been found that either increase or decrease repression their molecular mechanism is not yet understood. The *araO*$_2$ site involved in DNA looping binds AraC protein about 50-fold less tightly than the araI site. This is sufficiently weak binding that in the absence of looping to araI, the *araO*$_2$ site is not significantly occupied by AraC protein.

The co-operativity in binding between *araO*$_2$ and araI as measured with an in vivo footprinting system developed for the purpose provided good supporting data for the existence of looping in the ara system. Extensive helical-twist experiments have been done on the ara system. About 75 different spacings have been generated and the degree of repression in each measured. Sufficient data is available that by eye or by Fourier transform there is no question that the periodicity in the ability to repress expression oscillates with a period of a little over 11 base pairs, and is clearly not 10.5 base pairs per turn.

The dramatic difference in helical repeat from the oft-measured value of 10.5 base pairs per turn of linear DNA in vitro must result from the linking-number deficit of DNA in vivo. DNA extracted from bacteria has a linking-number deficit that is commonly measured as supercoiling. In reality, the linking-number deficit can be partitioned between unwinding the DNA, which increases the number of base pairs per turn, and supercoiling. Clever measurements show that in vivo the partitioning between the two is roughly half and half. At the physiological linking-number deficit of -0.06 and such a 50-50 partitioning, the helical repeat of the DNA would be about 10.8 base pairs per turn.

Therefore there remains 0.3 or so of base pairs per turn unaccounted for. It will be interesting when sufficiently accurate measurements have been made on other systems to see their helical repeat values.

AraC protein is a dimer in solution and binds to linear DNA as a dimer. Therefore it was expected that looping in the ara system would involve the binding of a dimer at araI to a dimer bound to araO_2, and it was a shock to discover that, in fact, just a single dimer of AraC protein generates the loops. When arabinose is added, the loop opens, contacts to araO_2 are lost, and DNA contacts are made only at aral This behaviour raises the question of what is it that makes the loop open and AraC protein now contact a larger stretch of DNA at araI? One attractive mechanism that could generate this behaviour is subunit orientation or position.

The subunits could be located in the absence of arabinose such that it is easier for them to contact two well-separated sites on the DNA via looping than two adjacent sites on the DNA. That is, the subunits could normally be misoriented. The addition of arabinose could alter the orientation such that the subunits become well positioned for binding to two adjacent sites on the DNA. This mechanism then could easily control looping, and could be extended to other systems in which looping is regulated.

Amongst the hints for looping in the lac system were experiments with chimeric *lac* repressor-β galactosidase molecules, the existence of multiple pseudo-operators, and the behaviour of operator mutants. In 1977 Kania & Muller-Hill reported the DNA-binding abilities of tetrameric hybrids of β-galactosidase consisting of the normal P-galactosidase subunit and β–galactosidase subunits containing the N-terminal 60-80 amino acids of lac repressor (37). The various types of hybrids were partially separated from one another and their binding to linear DNA was investigated with the filter-binding assay in which uncomplexed DNA freely passes through a filter, but the protein bound on the

DNA retards the DNA on the filter. Although significant cross-contamination existed in samples with differing numbers of DNA-binding domains, and there could be questions about proper positioning of DNA-binding domains or conformational effects on tetramers containing two or four of the repressor domains, a tighter DNA binding was seen in the fractions thought to contain molecules with four repressor domains than in the fractions enriched in hybrids containing two repressor domains.

The authors suggested that these results could result from the binding of repressor to two operators on the DNA simultaneously. The authors did not propose that the effect occurs in vivo, and this work was apparently not pursued between 1977 and 1986. In retrospect, it is difficult to understand any higher binding affinity seen by the tetrameric molecule because subsequent studies have shown that supercoiling is required for the natural operators to be able to engage in looping. Cooperativity in repressor binding to lac operators was also an overlooked clue to the existence of DNA looping. For some years, the behaviour of operator mutants in the lac operon was most puzzling.

Point mutations in the operator seemed to raise the basal level of expression of the operon at most only 20-fold, yet study of the binding affinities to the operators in vitro and consideration of the in vivo concentrations of lac repressor suggested that the mutations should have had a much greater effect on repression. Apparently DNA looping ameliorated the effects of the operator constitutive mutations. When wild-type lac operator was placed various distances from a mutant operator, it was apparent that the presence of the wild-type operator aided the binding of repressor to the mutant operator. By this time the concept of looping as a mechanism in gene regulation was accepted and the data was interpreted as possibly resulting from looping. Similarly, placing the lac operators upstream of the lac promoter gave no repression and placing an operator downstream gave moderate repression, but both operators together generated strong repression. Whether the repression results from looping, helping the repressor bind to the downstream site and thereby serve as a blockade to an active RNA polymerase, or whether the loop poses a steric hindrance to the binding of the polymerase is not yet known. An elegant experiment showed that looping in the *lac* system is a natural part of the operon's regulation. This made use of the two pseudo-operators found near the lac regulatory region.

Since 1974 it had been known that in addition to the primary lac operator located just downstream from the transcription initiation site, two additional weaker binding operators existed in the lac operon. For more than 10 years these were thought merely to be remnants of evolution. To investigate a role of the downstream operator O_2, which is located 481 base pairs downstream from the primary operator, eight bases in lac O_2 were altered in such a way as not to change the amino acids encoded by this region of the DNA.

Thus the β-galactosidase remained unaltered, but O_2 was so greatly altered that lac repressor no longer could bind. These changes were found to decrease the repression of the operon about fivefold. That is, they raised the uninduced or basal level fivefold, but had almost no effect on the induced activity of the promoter. This result indicates that the second lac operator does play a role in repressing the operon. Apparently DNA looping frequently occurs in vivo between the main lac operator and the downstream pseudo-operator. The second pseudo-operator is located 83 base pairs upstream from O_1, at the end of *lacl*. It also apparently plays a role in normal regulation of the operon, for its destruction raises the basal level of *lacZ* about threefold.

Nearly the full set of experiments that can demonstrate looping have now been applied to the lac operon, and with dramatic success. In vitro and in vivo footprinting has been used to show cooperativity in the binding of repressor to the main lac operator and the two pseudo-operators O_2, and O_3 located 83 and 481 base pairs from the operator. DNA-binding studies of lac repressor-operator interactions have been performed with the filter-binding assay.

The dissociation rate of repressor from such DNA is much reduced if the DNA is supercoiled and if it contains one of the pseudo-operators in addition to the operator at the promoter. In addition to using supercoiled DNA and the natural lac operators, looping in lac has also been studied by increasing the energies available for looping by using perfectly symmetric *lac* operators. Repressor binds to such operators about 100 times as tightly as it binds to the wild-type operator. With such operators, linear DNA could be forced to loop and at high repressor concentrations, "crossjoined" or sandwich structures could also be formed.

The structures could be detected by gel retardation or observed by electron microscopy. They were identified as looped structures by their enhanced stability and their absence when the two operators were rotated half a turn with respect to one another. Additionally, the presence of these loops on the DNA generated the characteristic alternating pattern of DNasel-hypersensitive and hyposensitive sites on the DNA between the two operators. In an effort to apply principles of statistical mechanics to looping in vivo in the lac operon, Mossing & Record examined operator spacings of 118, 185, and

283 base pairs and concluded that supercoiled DNA in vivo possessed the same stiffness and therefore experienced the same difficulty in bending sharply to form small loops as would be predicted from the hydrodynamically determined stiffness of linear DNA. Later it was learned that the inability of the closest spaced operators, 118 base pairs, to loop in their experiment apparently resulted not from stiffness of the DNA, but from the fact that they were on opposite faces of the DNA as the helical repeat of DNA in vivo is not 10.5 base pairs per turn as it is on linear DNA in vitro.

This they found in a subsequent helical-twist experiment in which the repression abilities of 12 different spacing mutations were clearly inconsistent with an helical repeat in vivo of 10.5 base pairs per turn. These authors have also considered the decrease in looping that occurs as loops become particularly large. This should occur when entropic effects might begin to become more important than the gains from not having to bend the DNA so sharply. This problem is paiticularly difficult, however, for in supercoiled DNA distal sites are forced into close contact with one another due to the wrapping, and simple polymer statistics should not be applied to the problem.

The *deo* system is an example where operator separations of up to 4000 base pairs showed only a small distance effect. An additional difficulty in attempting to calculate looping is the unknown flexibility of the protein. Undoubtedly some proteins possess significant flexibility and their bending will reduce the amount of bending required by the DNA to form a loop. DEO The protein products of the deo operon in Escherichia coli permit the cells to catabolize nucleosides. At least four enzymes are involved in the catabolism and three proteins are involved in the regulation. The synthesis of *deoC, deoA, deoB,* and *deoD* message initiates from two promoters separated by 599 base pairs and lying upstream of the four deo genes. Messenger ,synthesis is regulated by three proteins, the cyclic AMP receptor protein, which activates transcription, and the CytR and DeoR repressor proteins. In the absence of CytR protein, DeoR can repress expression of the Deo proteins about 60-fold. This repression requires three operator sites, one overlapping each of the -10 regions of the two promoters, and a third site lying an additional 270 base pairs upstream of the first two.

The integrity of all three operators is required for maximal repression, suggesting that a double DNA loop might form. Indeed, such is possible, for the molecular weight of DeoR protein based on gel exclusion chromatography indicates that the protein is an octamer, and therefore it could bind three, or possibly four of the 16-base-pair palindromic binding sites simultaneously. Double looped structures can be formed in vitro and observed by electron microscopy.

The potential energies available for looping of the deo system must be large, for loops will form on linear DNA, and particularly large loops can form and repress in vivo. When only two operators are present, one of which overlaps the -10 region of a promoter, repression falls from 23-fold to 3.6-fold to 3.0-fold as the spacing between the two operators increases from 567 base pairs to 3700 base pairs to 4685 base pairs.

Other looping systems

The gal operon of Escherichia coli is subject to a complex control as galactose can be catabolized when it is available, but also must always be present to fulfill needs for cell wall synthesis. Two promoters separated by about 10 base pairs serve the *galETK* genes. The activity of one is stimulated by cyclic AMP receptor protein, whereas the other is repressed by cyclic AMP receptor protein. The activity of both is repressed by GalR repressor, which binds both at positions -60 and at +55 with respect to the start of transcription. Instead of repressing via steric hindrance in which repressor bound to either of the operators blocks the binding of RNA polymerase or cyclic AMP receptor protein, a DNA loop formed by repressor bound at both sites generates a loop that represses the promoter.

Most likely the binding rate of RNA polymerase to such a sterically constrained promoter is much reduced from the nonlooped state. An elegant experiment showing that the repression was generated by looping was the finding that substitution of either binding site for the lac repressor binding site left the system poorly repressed, but when both of the gal operators were changed to lac operators, repression was strong when lac repressor was present. As expected, if a mutant lac repressor unable to form tetramers and therefore unable to loop was used, repression was poor. The genes involved in the uptake and metabolism of N-acetylglucosamine are controlled by the NagC protein, a repressor that loops. This may be a particularly good system for physical studies as short linear DNA fragments can be looped by the NagC protein.

Activator systems

A complex system utilizing a two-component signal-transducing system as well as protein phosphorylation regulates synthesis from the genes involved in nitrogen metabolism in Escherichia coli. The major difference between transcription from genes using the a 51 polymerase and those using 971 is that transcription from the former requires an additional activating protein. Transcription of the gln genes requires either the NtrC protein or the NifA protein, which bind to DNA one or two hundred base pairs away from the promoter (61) and form a loop in the DNA as they interact with RNA polymerase. Although no plausible mechanism except DNA looping could explain the action-at-a-distance effects shown by the activating protein NtrC in the transcription of glnALG genes, a nice looping confirmation was performed using linked or concatenated DNA circles as described earlier.

The electron microscopic evidence for looping in the *nif* system is rather inconclusive as only 5% of the DNA molecules with bound protein were observed to be looped. Undoubtedly NtrC and bound RNA polymerase can interact in the samples prepared for microscopy, but random aggregation of protein would produce the same result. The regulatory proteins NtrC and NifA regulate nitrogen fixation genes called nif in *Klebsella* pneumoniae, a close relative of Escherichia coli.

Helical-twist experiments in which 5, 11, 15, and 21 base pairs were added to the natural spacing of about 150 base pairs between the NtrC-binding site and the promoter for *nifLA* showed a cyclic dependence. Additional evidence for DNA looping in nif genes is that even if the natural spacing of one hundred or so base pairs from the promoter to the NifA-binding site were increased to as many as several thousand, activation by NifA could still be detected.

Eukaryotic enhancers in general

Eukaryotic enhancers activate and regulate transcription initiating at promoters located hundreds, thousands, and sometimes tens of thousands of base pairs away. Enhancers can be moved about and turned end for end, usually with only a small effect on their stimulatory powers. Enhancers were first recognized in animal viruses, and initially were hypothesized as being entry points for RNA polymerase molecules that would then drift to the promoter to start transcription.

One test of the drifting model is to place a blockade between the enhancer and promoter. Ptashne & Brent did this with the *gal4* system in yeast and found strong interference with promoter activity. This they interpreted as evidence in favour of the bind and drift model. After finding evidence for looping in vitro with the lambda repressor system, looping seemed more plausible and the blockade was interpreted as interfering with looping instead. Relatively few definitive experiments have been done to prove that proteins bound to enhancers generate loops as they interact with other parts of the transcription apparatus at the promoter.

Nonetheless, the prevailing opinion seems to be that most enhancers use a looping mechanism. This certainly seems the most plausible for enhancers with large separations between them and their promoters. Further, DNA looping is almost the only way that transvection in Drosophila could work. There an enhancer on one chromosome activates a promoter on a different but homologous chromosome. A few helical-twist experiments have been done with eukaryotic enhancers. Those that showed some helical effects were the late promoter of SV40, hand the heat-shock promoter of Drosophila.

Other promoters fail to show elical-twist effects. These negative findings can be the result of several factors. The energies available for loop formation may greatly exceed the energetic costs of torsionally twisting the DNA. Another reason may be that the proteins involved are sufficiently flexible that they can accommodate substantial misorientations of their binding sites. This is highly likely, for many of the eukaryotic enhancer binding proteins seem to consist of loosely connected domains that can freely be interchanged without loss of activity. As discussed earlier, tethering experiments support DNA looping. They have been done with the ribosomal promoter of Xenopus and the viral SV40 promoter . Electron microscopy of enhancer binding proteins binding together and to two sites on the DNA suggest, but hardly prove, the existence of DNA looping.

Looping in Genetic

Because the DNA sites participating in site-specific genetic recombination must be close to one another at the time of strand exchange, proteins bound to these sites bind to each other. Consequently DNA looping, or its equivalent if the sites are on two different DNA molecules, occurs. These types of reactions are found in the *hin* inversion system of Salmonella, the gin and pin inversion systems of phages Mu and P_1, resolution reactions of transposons, the transposition system of Mu, and the integration-excision system of lamb-doid phages.

The resolvase protein encoded by transposons such as Tn_3 and Mu phage catalyses an exchange reaction between two *res sites* in the same orientation on the same DNA molecule. When the sites are oppositely oriented there is no reaction or a greatly reduced reaction. Initially this fact was taken as evidence for a DNA-threading reaction by resolvase from one res site to the other. By this means the recombinase could determine the relative orientation of the two sites. It did not seem possible that the two sites could communicate their relative orientations by any other means. Closely reasoned topological experiments followed by an elegant linked-circle experiment with the Mu phage transposition system however, changed thinking on this issue.

DNA molecules with two properly oriented Mu *att* sites that were capable of recombining were used in this experiment. The DNA substrates for this experiment also contained lambda phage attB and attP sites oriented such Fig. 3. that the integration reaction between them would either invert one of the Mu att sites or separate the two att sites onto two separate DNA molecules. Once the DNA was supercoiled and capable of supporting a reaction between the att sites, the lambda phage integrase reaction between the sites, which either inverted an att site or placed the two sites on two different molecules, had no effect on the reaction.

The explanation for these results is that the two att sites are placed in close proximity and in the proper spatial orientation via the supercoiling of the DNA. If this supercoiling persists while the DNA molecule is cut and rejoined at sites elsewhere, the reaction between the att sites is unaltered. The cutting and rejoining by the lambda *int* system can separate the sites or even invert them topologically. The lambda phage integration-excision structure itself can also be thought of as a loop.

The complex of Int, IHF, and Xis proteins at the phage attachment region is sharply bent by the proteins and appears to be nearly looped itself. Of course the Int protein participates in the strand scission and rejoining reactions, but it also helps hold the DNA looped by binding both at the crossover region as well as 125 and 150 base pairs away.

Although six dimers; of proteins, including three dimers; of IHF, are present in the structure, they appear not to require direct protein-protein interaction. The cyclic AMP receptor protein bends DNA about as much as IHF protein, and it can replace IHF protein in the intasome. Also, a sequence of DNA possessing oriented runs of As and Ts so that it naturally bends about 100 degrees can replace one of the IHF proteins.

Looping in regulation of DNA replication

DNA replication is a highly important process, and it is likely that as discussed earlier, proteins that bind to DNA and that interact are able to form loops in the DNA. As a collection of proteins binds to the origin of replication of phage lambda and *E.* coli, it seems reasonable that the activity of most origins will require binding by multiple proteins. Such a situation opens the possibility of DNA looping, and indeed looping between two lambda replication origins that had been placed on the same DNA molecule mediated by the lambda O protein has been observed. It remains to be seen whether this type of looping is of biological significance.

DNA Libraries

DNA from the genome of an organism of interest and cloning those regions in suitable vectors. We can take all the DNA of an organism and clone it in chunks of reasonable size. We can do this through DNA libraries. Consider the case of constructing a library of the human genome. There are six billion base pairs in a diploid human cell. If we consider that 20,000 base pairs is a reasonable size for a cloned insert, we will need a minimum of 300,000 different recombinant DNAs.

It is quite a feat to achieve this number of different recombinants, and in practice we would need several times this minimum to ensure full representation and to account for the vector molecules that do not acquire an insert. To separate the individual members of the population of plasmid DNA molecules by cloning them is the second step. One way to simplify the procedure is to take advantage of the fact that widely used bacterial plasmids frequently carry genes for antibiotic resistance and for enzymes that catalyse reactions to give rise to products with a noticeable colour. In our example, the plasmid has a gene for ampicillin resistance and a gene for an enzyme (β-galactosidase) that catalyses the conversion of a colourless derivative of the sugar galactose to a compound with an easily visible blue colour.

The site for cleavage by the restriction endonuclease to be used for construction of the recombinant DNA lies within the gene for β-galactosidase. If recombination takes place, the gene for β-galactosidase will no longer be functional. The appearance or nonappearance of a blue colour and the presence or absence of antibiotic resistance will be useful in detecting the presence of plasmids and whether they have acquired a DNA insert. After preparing population of plasmids, it is mixed with bacteria that are ampicillin sensitive and grown in a medium containing the galactose derivative.

Cells that do not acquire a plasmid will not grow on a medium that contains ampicillin, nor will they produce a blue colour. Cells that acquire a plasmid without a DNA insert will grow on a medium that contains ampicillin, because each of these cells has an intact gene for antibiotic resistance, and they will produce blue colonies because the β-galactosidase gene is also intact. Cells that acquire a plasmid with an insert will be resistant to ampicillin and grow on the ampicillin-containing medium but will not produce a blue colour. The group of clones that has acquired a plasmid with an insert constitutes the recombinant library. The whole library can be stored for future use, or a single clone can be selected for further study.

The process of constructing a DNA library can be quite laborious, leading many researchers to obtain previously constructed libraries from other laboratories. Some journals require that libraries and individual clones that have been discussed in articles published by them should be freely available to other laboratories. RNA libraries are not constructed and cloned as such. Rather, the RNA of interest (usually mRNA) is used as the template for synthesis of complementary DNA (cDNA) in a reaction catalysed by reverse transcriptase.

The cDNA is incorporated into a vector. Ligating the cDNA to a vector requires use of a synthetic linker. The process of producing a cDNA library is virtually identical to that for constructing a genomic DNA library. After a DNA library has been constructed, if someone wants to find a single clone, out of the hundreds of thousands, or possibly millions, in the library. This degree of selectivity requires specialized techniques.

One of the most useful of these techniques depends on separation and annealing of complementary strands. An imprint of the dish on which the bacterial colonies have grown is taken by blotting. A nitrocellulose disc is placed on the dish and then removed. Some of each colony or plaque is transferred to the disc, and the position of each is the same as it was on the dish. The rest of the original colonies or plaques remain on the dish and can be stored for future use. The nitrocellulose disc is treated with a denaturing agent to unwind all the DNA on it. The next step is to expose the disc to a solution that contains single-stranded DNA (or RNA) that has a sequence complementary to one of the strands in the clone of interest.

The added single-stranded nucleic acid is called the probe; it is tagged in some way, with a radioactive label or with a luminescent compound, to ensure detection. The probe anneals to the DNA of interest and only to that DNA. Any excess solution is washed off the nitrocellulose disc. In the case of radioactive probes, the disc is placed in contact with x-ray film. Only those spots on the disc in which some of the probe has annealed to the DNA already there are radioactive, and only those spots expose the x-ray film. Nonradioactive probes depend on the use of compounds, such as biotin, that react to emit light. In either case, the probe causes the desired DNA spot to "light up."

Since the original Petri dish has been saved, the desired clone can be picked off the plate and allowed to reproduce. If the nucleotide sequence of the desired DNA segment is not known and no probe is available, a complication arises. If the gene of interest directs the synthesis of a given protein, one chooses a vector that will allow cloned genes to be transcribed and translated. If the presence of the desired protein can be detected by its function, that serves as the basis for detecting it.

Polymerase chain reaction

To increase the amount of a given DNA many times over without cloning that DNA, the polymerase chain reaction (PCR) is used. Any chosen DNA can be amplified, and it need not be separated from the rest of the DNA in a sample before the procedure is applied. PCR copies both complementary strands of the desired DNA sequence. At the start of the process, the two strands are separated by heating, after which short oligonucleotide segments are added in large excess and, with cooling, are allowed to anneal to the desired DNA.

These short segments are complementary to the ends of the DNA chosen for amplification and are there to serve as primers for copying the entire complementary strand in the presence of DNA polymerase and the other factors needed for the synthesis. The two complementary strands grow in the 5′ to 3′ direction. This first round doubles the amount of the desired DNA. The process of unwinding the two strands, annealing the initiators, and copying the complementary strand is repeated, bringing about a second doubling of the selected double stranded DNA. It is not necessary to add more primer, since it is present in large excess. It is also not necessary to add more DNA polymerase.

A heat-stable polymerase from *Thermus aquaticus* remains active under the conditions of amplification. The whole process is automated. Control of the temperature to which the strands are heated to separate them is crucial, as is the temperature chosen for annealing the primers. The time allowed for heating and annealing is an important factor, as well. The amount of DNA continues to double in subsequent rounds of amplification. After about an hour, and 25 to 40 cycles of replication, millions to hundreds of millions of copies of the desired DNA segment, usually a few hundred to a few thousand base pairs in length, are obtained. Other DNA sequences are not amplified and do not interfere with the reaction or subsequent use of the amplified DNA. Amplification of the amounts of DNA in extremely small samples has made it possible to obtain accurate analyses that were not possible earlier. Forensic applications of the technique have resulted in positive identifications of crime victims and suspects.

Gene mapping

Genetic map or linkage map of an organism is the order of genes on a chromosome, an order that can be determined by the methods of classical genetics. As an example, consider the case bacterium *Escherichia coli.* It can transfer DNA from one cell to another by a process known as conjugation. The cells are in physical contact as DNA is transferred from one to another. The longer the cells are in contact, the more DNA can be transferred. Genes close to each other on the bacterial genome travel together in the course of conjugation. If the flask containing the cells is shaken, the contact is broken and less of the DNA is transferred.

The amount of DNA transferred in the process can be detected by use of recipient cells that are missing some genes and testing for the presence of those genes in the recombinants. The order in which genes are transferred can be determined by varying the amount of time allowed for conjugation. Genes that are closer together tend to be linked in the course of recombination and to be transferred together even during short recombination periods. Genes that are farther apart are much less likely to be linked and thus are less likely to be transferred together. This method and variations on it produce the genetic map. The construction of the map is based entirely on genetic methods that rely on visible traits and does not deal directly with the nature of the molecules involved. It is also possible to construct a physical or molecular map of the DNA sequence on chromosomes.

Construction of a physical map of chromosomes depends in an important way on the organism involved. In a prokaryote there is normally a single chromosome not separated from the rest of the cell; a prokaryotic chromosome is a circle of double-helical DNA. In a eukaryote there are multiple chromosomes, sequestered in the nucleus; each contains a single, linear DNA double helix. In prokaryotes, genes lie close to each other without intervening DNA. In eukaryotes, genes are separated by long stretches of noncoding DNA sequences along the chromosome. Dealing with all these complexities requires the powerful methods developed for recombinant DNA technology. Physical map of the *E. coli* genome was obtained by analysing the restriction fragments that are produced when the entire genome is digested with a restriction endonuclease.

A particular enzyme was chosen that gives rise to relatively few fragments, which can be separated by gel electrophoresis. DNA from the gel is transferred to a nitrocellulose filter in the manner that we saw with DNA transfer in analysis of DNA libraries. The positions of the DNA fragments on the nitrocellulose filter exactly match the positions of the DNA fragments in the original gel. A probe solution containing single-stranded DNA of a given sequence is incubated with the nitrocellulose filter. The probe DNA binds only to those fragments that contain the sequence of interest. Detection is done by autoradiography or by chemiluminescence, depending on the way in which the probe DNA is labeled. In this manner, it is possible to determine which restriction fragments contain all or part of a given gene. This technique is called DNA blotting.

Positions of the fragments on the circular genome were determined by comparison with the classical genetic map, on which the positions of more than 1000 genes had already been determined at the time this work was done. Cloned segments of these known genes were available as probes in DNA blotting to locate a given gene on a given fragment. Classical genetic studies on the fruit fly Drosophila melanogaster in the early 20th century determined linkage patterns of genes using methods similar to those just described for bacteria. This work produced detailed maps of genes on chromosomes, but the positions of many more genes were completely unknown. Even less information was available for other eukaryotes, and this situation was unchanged for many years. More recently it has become possible to obtain genetic maps of whole genomes of bacteria and yeast and to launch the Human Genome Project, which is currently mapping the entire human genome. In humans, the first genes to be mapped were those on the *X* chromosome.

Females have two *X* chromosomes, whereas males have one *X* chromosome and one Y chromosome. If a gene on the single *X* chromosome of a male is nonfunctional, there is no chance of the function's being replaced by the corresponding gene on the other *X* chromosome, as can happen with females. As a result, some human males exhibit traits that arise from nonfunctional genes on the *X* chromosome. Examples include red-green colour blindness and diseases such as hemophilia and Duchenne muscular dystrophy.

Females are very seldom affected by these traits but are carriers of them; a mother passes them on to her sons. Such observations made it possible to locate on the X chromosome the genes involved in the traits. Progress in locating human genes on chromosomes other than the sex chromosomes (*X* and *Y*) was slow until recombinant DNA methods became available. The family of genes that encodes the β-chain of hemoglobin consists of a cluster arranged along chromosome II in the order in which the genes are expressed in the course of development.

Five different β-type chains are made from the early embryonic stage to shortly after birth. The form known as ε-globin is made only in the early embryo stage. The two γ–globin forms, Gγ and Aγ, are made later in fetal life, while the gene for the epsilon form is switched off. After birth, the genes for δ-globin and β–globin are expressed, and the genes for the gamma forms are switched off in turn. Note that the map also contains two pseudogenes, designated as ψβ.

Pseudogenes are nonfunctional copies of genes that cannot normally be expressed because of some mutation, frequently a deletion. Pseudogenes occur to varying extents in eukaryotes and are particularly prevalent in mammals. A similar arrangement appears in the α-globin gene cluster on chromosome 16. The variant of α-globin that is produced in the early embryo is ξ-globin. This gene is switched off later in development, at which time the α-globin gene is expressed. Note the presence of pseudogenes for both the alpha and zeta forms (ψα and ψξ respectively). The mechanism of regulation of these two gene clusters on different chromosomes is a topic of considerable interest to biologists. It is not surprising that human globin genes were among the first mammalian genes to be cloned using recombinant DNA methods.

There was considerable incentive to develop gene therapies for the genetic diseases that arise from abnormal hemoglobins; it was also comparatively easy to carry out the actual cloning process. Protein synthesis in human red blood cells is largely dedicated to the production of hemoglobin, which accounts for about 90 per cent of the protein in these cells. Most of the mRNA in red blood cells is of two kinds, those for the α-globins and those for the β-globins. Using these two readily available mRNAs, it was possible to construct probes to identify the globin genes in genomic libraries. Each of these genes was contained in a clone that contained stretches of cDNA flanking the gene itself.

The flanking sequences were cloned by themselves; they could then be used to construct probes for segments of DNA farther along the chromosome. This process is called chromosome walking. It requires overlapping fragments derived from the genomic library by use of several restriction endonucleases. A portion of a cloned restriction fragment can be subcloned and used as a probe for an overlapping sequence in the genomic library. A portion of this sequence can be subcloned in turn and used as a probe to rescreen the genomic library. The process is repeated until a gene of interest is reached. In organisms with two sets of chromosomes, a given gene on one chromosome may differ slightly from the corresponding gene on the paired chromosome. In the language of genetics, these genes are alleles. When they are the same on the paired chromosomes, the organism is homozygous for that gene; when they differ, the organism is heterozygous.

A difference between alleles, even a change in one base pair, can mean that one allele has a recognition site for a restriction endonuclease and the other does not. Restriction fragments of different sizes are obtained on treatment with the nuclease; they are called restriction fragment length polymorphisms, or RFLPs (pronounced "riflips") for short. These polymorphisms are analysed using gel electrophoresis to separate the fragments by size, followed by blotting and the annealing of a probe for a gene being sought. Such polymorphisms are quite common, much more so than the traits such as eye colour and inherited diseases that were used for earlier genetic mapping.

RFLPs can be used as markers on the genetic linkage map in the same way as mutations in visible traits, because they are inherited in the manner predicted by classical genetics. However, since they are more abundant than mutations within coding sequences, they have provided many more markers for detailed genetic mapping. RFLP analysis was used extensively in the process of locating the altered gene that causes cystic fibrosis, a prevalent genetic disease.

Once the gene was located on chromosome 7, a series of RFLP markers was used to help map its exact position by constructing a linkage map based on the RFLP positions relative to the cystic fibrosis gene. Then the gene was isolated from restriction endonuclease digests, cloned, and its protein product characterized. The protein in question is involved in the transport of chloride ion (Cl^-) through membranes. If this protein is defective, chloride ions remain in the cells and take up water by osmosis from the surrounding mucus.

The mucus thickens as a result. In the lungs the thickened mucus favours infections, particularly pneumonia. The results of this disease can be tragic, with affected children seldom living past their early twenties. This information deepens our insight into the nature of cystic fibrosis and provides approaches to new treatment. The primary structure of a protein determines its secondary and tertiary structures.

The same is trite of nucleic acids; the nature and order of monomer units determine the properties of the whole molecule. The base pairing in both RNA and DNA depends on a series of complementary bases, whether these bases are on different polynucleotide strands, as in DNA, or on the same strand, as is frequently the case in RNA. The fact that the number of commonly occurring bases in nucleic acids is less than the number of amino acids in proteins has made the determination of the primary structure of nucleic acids a challenging task.

The situation is analogous to a jigsaw puzzle in which all the pieces are the same colour, which is more challenging than a jigsaw puzzle with a coloured pattern that helps to relate the pieces. Sequencing of nucleic acids, is now fairly routine, and this relative case would have amazed the scientists of the 1950s and 1960s. The method devised by Sanger and Coulson for determining the base sequences of nucleic acids depends on selective interruption of oligonucleotide synthesis. A single-stranded DNA fragment whose sequence is to be determined is used as a template for the synthesis of a complementary strand. The new strand grows from the 5′ to the 3′ end. This unique direction of growth is true for all nucleic acid synthesis. The synthesis is interrupted at every possible site in a population of molecules. The interruption of synthesis depends on the presence of 2′,3′-dideoxyribonucleoside triphosphates (ddNTPs).

The T-hydroxyl group of deoxyribonucleoside triphosphates (the usual monomer unit for DNA synthesis) has been replaced by a hydrogen. These ddNTPs can be incorporated in a growing DNA chain, but they lack a 3'-hydroxyl group to form a bond to another nucleoside triphosphate. The incorporation of a ddNTP into the growing chain causes termination at that point. The presence of small amounts of ddNTPs in a replicating mixture causes random termination of chain growth. DNA to be sequenced is mixed with a short oligonucleotide that serves as a primer for synthesis of the complementary strand.

The primer is hydrogen bonded to the 3'end of the DNA to be sequenced. The DNA with primer is divided into four separate reaction mixtures. Each reaction mixture contains all four deoxyribonucleoside triphosphates (dNTPs), one of which is labeled to allow the newly synthesized fragments to be visualized by autoradiography or by fluorescence. In addition, each of the reaction mixtures contains one of the four ddNTPs. Synthesis of the chain is allowed to proceed in each of the four reaction mixtures. In each mixture, chain termination occurs at all possible sites for that nucleotide. When gel electrophoresis is performed on each reaction mixture, a band corresponding to each position of chain termination appears. The sequence of the newly formed strand, which is complementary to that of the template DNA, can be "read" directly from the sequencing gel. A variation on this method is to use a single reaction mixture with a different fluorescent label on each of the four ddNTPs.

Chapter 6

Amino Acids and Peptide Bonds

The α-carbon is also bonded to a hydrogen and to the side chain group, which is represented by the letter R. The R group determines the identity of the particular amino acid (Figure 1). The two-dimensional formula shown here can only partially convey the common structure of amino acids, because one of the most important properties of these compounds is their three-dimensional shape, or stereochemistry.

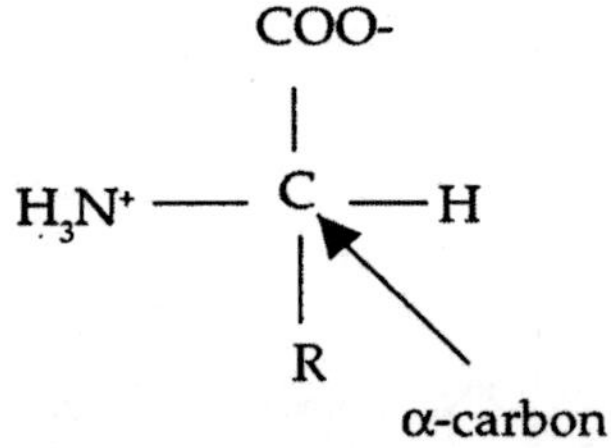

Fig. 1. The general formula of amino acids, showing the ionic forms that predominate at pH 7.

Many pairs of objects that are mirror images can be superimposed on each other; two identical solid-coloured coffee mugs are an example. In other cases the mirror-image objects cannot be superimposed on one another but are related to each other as the right hand is to the left. Such nonsuperimposable mirror images are said to be chiral (from the Greek cheir, "hand"); many important biomolecules are chiral. A frequently encountered chiral center in biomolecules is a carbon atom with four different groups bonded to it. Such a center occurs in all amino acids except glycine.

Glycine has two hydrogen atoms bonded to the a-carbon; in other words, the side chain (R group) of glycine is hydrogen. Glycine is not chiral (or, alternatively, is achiral) because of this symmetry. In all the other commonly occurring amino acids the cc-carbon has four different groups bonded to it, giving rise to two nonsuperimposable mirror-image forms.

The dashed wedges represent bonds directed away from the observer, and the solid tri-angles represent bonds directed out of the plane of the paper in the direction of the observer. The two possible stereoisomers of another chiral compound, L- and D-glyceraldehyde, are shown for comparison with the corresponding forms of alanine. These two forms of glyceraldehyde are the basis of the classification of amino acids into L and D forms. The terminology comes from the Latin laevus and dexter, meaning "left" and "right," respectively.

The two stereoisomers of each amino acid are designated as L and D amino acids on the basis of their similarity to the glyceraldehyde standard. In the L, form of glyceraldehyde, the hydroxyl group is on the left side of the molecule, and in tile D form it is on the right side. In an amino acid, the position of the amino group on the left or right side of the α-carbon determines the L, or D designation.

The amino acids that occur in proteins are all of the L form. Although D amino acids occur in nature, most often in bacterial cell walls and in some antibiotics, they are not found in proteins.

Properties of amino acids

The R groups, and thus the individual amino acids, are classified according to several criteria, two of which are particularly important. The first of these is the polar or nonpolar nature of the side chain. The second depends on the presence of an acidic or basic group in the side chain. Other useful criteria include the presence of functional groups in the side chains and the nature of those groups. As mentioned, the side "chain" of the simplest amino acid, glycine, is a hydrogen atom, and in this case alone two hydrogen atoms are bonded to the α-carbon.

In all other amino acids the side chain is larger and more complex. Side-chain carbon atoms are designated with letters of the Greek alphabet, counting from the α-carbon. These carbon atoms are, in turn, the β–, γ–, δ–, and ε-carbons; a terminal carbon atom is referred to as the (ω-carbon, from the name of the last letter of the Greek alphabet. We frequently refer to amino acids by three-letter or one-letter abbreviations of their names. Table 1 lists these abbreviations.

Group 1: Amino acids with nonpolar side

Chains

One group of amino acids has nonpolar side chains. This group consists of alanine, valine, leucine, isoleucine, proline, phenylalanine, tryptophan, and methionine. (In some classification schemes, glycine is placed in this group because it does not have a polar side chain.) In several members of this group - namely alanine, valine, leucine, and isoleucine-each side chain is an aliphatic hydrocarbon group. (In organic chemistry, the term "aliphatic" refers to the absence of a benzene ring or related structure.)

Table 1. Names and abbreviations of the common amino acids

Amino acid	*Three-letter abbreviation*	*One-letter abbreviation*
Alanine	Ala	A
Arginine	Arg	R
Asparagine	Asn	N
Aspartic acid	Asp	D
Cysteine	Cys	C
Glutamic acid	Glu	E
Glutamine	Gln	Q
Glycine	Gly	G
Histidine	His	H
Isoleucine	Ile	I
Leucine	Leu	L
Lysine	Lys	K
Methionine	Met	M
Phenylalanine	Phe	F
Proline	Pro	P
Serine	Ser	S
Threonine	Thr	T
Tryptophan	Trp	W
Tyrosine	Tyr	Y
Valine	Val	V

Proline has an aliphatic cyclic structure and the nitrogen is bonded to two carbon atoms. In the terminology of organic chemistry, the amino group of proline is a secondary amine. In contrast, the amino groups of all the other common amino acids are primary amines. In phenylalanine the hydrocarbon group is aromatic (contains a cyclic group similar to a benzene ring) rather than aliphatic. In tryptophan the side chain contains an indole ring, which is also aromatic. In methionine the side chain contains a sulfur atom in addition to aliphatic hydrocarbon groupings.

Group 2: Amino acids with electrically

Neutral polar side chains

Glycine is also included here for convenience because it lacks a nonpolar side chain, but, as mentioned be fore, it is also put in Group I by some biochemists because the C–H bond is nonpolar. In serine and threonine the polar group is a hydroxyl (-OH) bonded to aliphatic hydrocarbon groups.

The hydroxyl group in tyrosine is bonded to an, aromatic hydrocarbon group, which does eventually lose a proton at higher pH. (The hydroxyl group in tyrosine is a phenol, which is a stronger acid than an aliphatic alcohol. As a result the side chain of tyrosine can lose a proton, whereas those of serine and threonine do not.) In cysteine the polar side chain consists of an -SH (thiol) group, which can react with other cysteine -SH groups to form disulfide (-S-S-) bridges in proteins in an oxidation reaction.

The thiol group can also lose a proton. The amino acids - glutamine and asparagine have amide groups, which are derived from carboxyl groups, in their side chains. Amide bonds do not ionize in the range of pH usually encountered in biochemistry. Glutamine and asparagine can be considered as derivatives of the Group 3 amino acids, glutamic acid and aspartic acid, respectively; those two amino acids have carboxyl groups in their side chains.

Group 3: Amino acids with carboxyl groups

Two amino acids, glutamic acid and aspartic acid, have carboxyl groups in their side chains in addition to the one present in all amino acids. A carboxyl group can lose a proton, forming the corresponding carboxylate anion, glutamate and aspartate, respectively, in the case of these two amino acids. Because of the presence of the carboxylate, the side chains of these two amino acids are negatively charged at neutral pH.

Glutamate and aspartate differ only by a $-CH_2-$ group in the side chain. The side-chain carbonyls form side-chain amide groups with $-NH_2$, yielding the analogous amino acids glutamine and asparagine. The side-chain amide groups are electrically neutral at neutral pH. For that reason, we include asparagine and glutamine in Group 2.

Group 4: Amino acids with basic side

Chains

There are three amino acids (histidine, lysine, and arginine) that have basic side chains. In lysine and arginine side chains are positively charged at neutral pH. In histidine the pK_a of the side-chain imidazole group is not far from physiological pH, and the properties of many proteins depend on whether individual histidine residues are or are not charged. In lysine the side-chain amino group is attached to an aliphatic hydrocarbon tail. In arginine the side-chain basic group, the guanidino group, is more complex in structure than the amino group, but it is also bonded to an aliphatic hydrocarbon tail.

Uncommon amino acids

Many other amino acids are known to exist, in addition to the ones listed above. Here are some examples out of many possibilities. Several amino acids are known to occur in some, but by no means all, proteins. They are derived from the common amino acids and are produced by modification of the parent amino acid after the protein is synthesized by the organism in a process called posttranslational modification. Hydroxyproline and hydroxylysine differ from the parent amino acids in having hydroxyl groups on their side chains; they are found only in a few connective tissue proteins, such as collagen.

Thyroxine differs from tyrosine in having an extra iodine-containing aromatic group on the side chain; it is found only in the thyroid gland, bound to the protein thyroglobulin.

Titration curves of the amino acids

In a free amino acid the carboxyl group and amino group of the general structure are charged at neutral pH - the carboxylate portion negatively and the amino group positively. Amino acids without charged groups on their side chains exist in neutral solution as zwitterions with no net charge. A zwitterion has equal positive and negative charges; in solution it is electrically neutral. Neutral amino acids do not exist in the form NH_2-CHR-COOH, that is, without charged groups. When an amino acid is titrated, its titration curve indicates the reaction of each functional group with hydrogen ion. In alanine, the carboxyl and amino groups are the two titratable groups. At very low pH, alanine has a protonated (and thus uncharged) carboxyl group and a positively charged amino group that is also protonated. Under these conditions the alanine has a net positive charge of 1. As base is added, the carboxyl group loses its pro ton to become a negatively charged carboxylate group, and the pH of the solution increases. Alanine now has no net charge. As the pH increases still further with addition of more base, the protonated amino group (a weak acid) loses its proton, and the alanine molecule now has a negative charge of 1.

The titration curve of alanine is that of a diprotic acid. The imidazole side chain also contributes a titratable group. At very low pH values, the histidine molecule has a net positive charge of 2, because both the imidazole and amino groups have positive charges. As base is added and the pH increases, the carboxyl group loses a proton to become a carboxylate as before, and the histidine now has a positive charge of I. As still more base is added, the charged imidazole group loses its proton, and this is the point at which the histicline has no net charge.

At still higher values of pH the amino group loses its proton, as was the case with alanine, and the histicline molecule now has a negative charge of 1. The titration curve of histidine is that of a triprotic acid. The amino acids have characteristic values for the K_as and pK_as of their titratable groups. The pK_as of (α-carboxyl groups are fairly low, around 2. The pK_as of amino groups are much higher, with values ranging from 9 to 10.5. The pK_as of side-chain groups, including side-chain carboxyl and amino groups, depend on the groups' chemical nature. Table 1 lists the pK_as of the titratable groups of the amino acids. The classification of an amino acid as acidic or basic depends on the pK_a of the side chain.

Table 2. pKa values of commons amino acids

Acid	COOH	α NH^+_3	α RH o: RH^+
Gly	2.34	9.60	
Ala	2.34	9.69	
Val	2.32	9.62	
Leu	2.36	9.68	
Ile	2.36	9.68	
Ser	2.21	9.15	
Thr	2.63	10.43	
Met	2.28	9.21	
Phe	1.83	9.13	
TrP	2.38	9.39	
Asn	2.02	8.80	
Gln	2.17	9.13	
Pro	1.99	10.6	
Asp	iog	9.82	3.86
Glu	2.19	9.67	4.25
His	1.82	9.17	6.0
Cys	1.71	10.78	8.33
Tyr	-2.20	9.11	10.07
Lys	2.18	8.95	16.53
Arg	2.17	9.04	12-.48

These groups can still be titrated after the amino acid is incorporated into a peptide or protein, but the pK_a of the titratable group on the side chain is not necessarily the same in a protein as it is in a free amino acid. In fact, it can be very different. For example, a pK_a of 9 has been reported for an aspartate side chain in the protein thioredoxin. The fact that amino acids, peptides, and proteins have different pKas gives rise to the possibility that they can have different charges at a given pH.

Alanine and histidine, for example, both have net charges of -1 at high pH, above 10; the only charged group is the carboxylate anion. At lower pH, around 5, alanine is a zwitterion with no net charge, but histidine has a net charge of +1 at this pH because the imidazole group is protonated. This property is the basis of electrophoresis, a method for separating molecules on the basis of charge. This method is extremely useful in determining the structure of proteins and nucleic acids. The pH at which a molecule has no net charge is called the isoelectric pH (given the symbol pI). At its isoelectric pH, a molecule will not migrate in an electric field. This property can be put to use in separation methods. The pI of an amino acid can be calculated by the following equation

$$pI = \frac{pK_{a1} + pK_{a2}}{2}$$

Peptide bond

The bond is formed between the α-carboxyl group of one amino acid and the α-amino group of the next one. Water is eliminated in the process, and the linked amino acid residues remain after the elimination of water. A bond formed in this way is called a peptide bond. Peptides are compounds formed by the linking of small numbers of amino acids, ranging from two to several dozen. In a protein, many amino acids (usually more than a hundred) are linked by peptide bonds to form a polypeptide chain. Another name for a compound formed by the reaction between an amino group and a carboxyl group is an amide. The carbon-nitrogen bond formed when two amino acids are linked in a peptide bond is usually written as a single bond, with one pair of electrons shared between the two atoms. With a simple shift in the position of a pair of electrons, it is quite possible to write this bond as a double bond. This shifting of electrons is well known in organic chemistry and results in resonance structures, structures that differ from one another only in the positioning of electrons. The positions of double and single bonds in one resonance structure are different from their positions in another resonance structure of the same compound. No single resonance structure actually represents the bonding in the compound, but all resonance structures contribute to the bonding situation.

The peptide bond can be written as a resonance hybrid of two structures, one with a single bond between the carbon and nitrogen and the other with a double bond between the carbon and nitrogen. The peptide bond has partial double bond character, and as a result the peptide group that forms the link between the two amino acids is planar.

The peptide bond is also stronger than an ordinary single bond because of this resonance stabilisation. There is free rotation around the bonds between the (α-carbon of a given amino acid residue and the amino nitrogen and carbonyl carbon of that residue, but there is no significant rotation around the pepticle bond. This stereochemical constraint plays an important role in determining how the protein backbone can fold.

Small peptides of physiological interest

The simplest possible covalently bonded combination of amino acids is a dipeptide, in which two amino acid residues are linked by a peptide bond. An example of a naturally occurring dipeptide is carnosine, which is found in muscle tissue. This compound, which has the alternative name β–alanyl-L-histidine, has an interesting structural feature.

The N-terminal amino acid residue, β-alanine, is structurally different from the (α-amino acids we have seen up to now. As the name implies, the amino group is bonded to the second or β–carbon of the alanine. The peptide bond in this dipeptide is formed between the carboxyl group of the β-alanine and the amino group of the histidine, which is the C-terminal amino acid. Glutathione is a commonly occurring tripeptide; it has considerable physiological importance because it is a scavenger for oxidizing agents.

Oxidation is the loss of electrons; an oxidizing agent causes another substance to lose electrons. (It is thought that some oxidizing agents are harmful to organisms and play a role in the development of cancer.) In terms of its amino acid composition and bonding order, it is γ–glutamyl-L-cysteinylglycine. Structure.

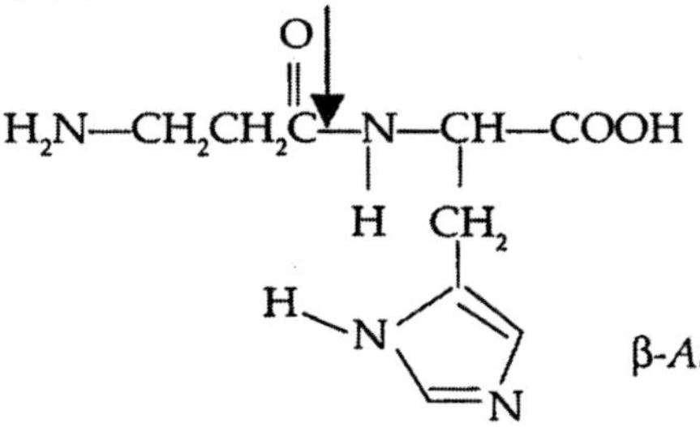

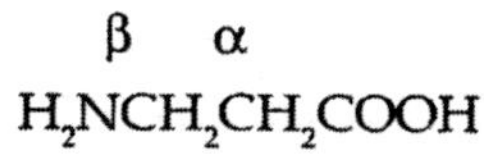

β-Alanine

β-Alanyl-L-histidine (carnosine)

Fig. 2. Structures of carnosine and its component amino acid.

The letter γ is the third letter in the Greek alphabet; in this notation it refers to the third carbon atom in the molecule, counting the one bonded to the amino group as the first. Once again the N-terminal amino acid is given first. In this case the γ-carboxyl group (the side-chain carboxyl group) of the glutamic acid is involved in the peptide bond; the amino group of the cysteine is bonded to it.

The carboxyl group of the cysteine is bonded, in turn, to the amino group of the glycine. The carboxyl group of the glycine forms the other end of the molecule, the C-terminal end. The glutathione molecule scavenges oxidizing agents by reacting with them. The oxidized form of glutathione is generated from two molecules of the reduced peptide by formation of a disulfide bond between the -SH groups of the two cysteine residues. Two pentapeptides found in the brain are enkephalins, naturally occurring analgesics (pain relievers). For molecules of this size, abbreviations for the amino acids are more convenient than structural formulas.

The two peptides in question, leucine and methionine enkephalin, differ only in their C-terminal amino acids. It is thought that the aromatic side chains of tyrosine and phenylalanine in these peptides play a role in their activities. It is also thought that there are similarities between the three-dimensional structures of opiates, such as morphine, and those of the enkephalins. As a result of these structural similarities, opiates bind to the receptors in the brain intended for the enkephalins and thus produce their physiological activities. Some important peptides have cyclic structures. Two well-known examples with many structural features in common are oxytocin and vasopressin. In each there is an -S-S- bond similar to that in the oxidized form of glutathione; the disulfide bond is responsible for the cyclic structure. Each of these peptides contains nine amino acid residues, each has an amide group, rather than a free carboxyl group, at the C-terminal end, and each has a disulfide link between cysteine residues at positions 1 and 6.

The difference between these two peptides is that oxytocin has an isoleucine residue at position 3 and a leucine residue at position 8, and vasopressin has a phenylalanine residue at position 3 and an arginine residue at position 8. Both of these peptides have considerable physiological importance as hormones.

Chapter 7

Biochemistry of Peroxisomes

The involvement of peroxisomes in ether lipid biosynthesis has been further substantiated, and evidence for a role in cholesteroland dolichol metabolism has been presented. The early history of the mammalian peroxisome and its relationship with microbody structures in other kingdoms has been reviewed extensively. Reviews concentrating on microbodies in plants, protozoa, and fungi are available. It is now well-documented that microbodies from these various sources contain different arrays of enzymes. More than a decade ago Tolbert presented a comprehensive summary in this series of the enzyme content of mammalian peroxisomes and their counterpart from the plant kingdom, glyoxisomes.

There have been many new developments since that time. The recognition of a new class of inheritable diseases in humans caused by the deficiency of peroxisomes or single peroxisomal enzymes has provided new insights into the function of processes taking place in peroxisomes or has assigned processes to peroxisomes that were previously thought to take place in other organelles.

Major breakthroughs in establishing the site of synthesis of peroxisomal proteins, their import into the organelle, and the targeting sequences involved have been reported. As a consequence of these studies, the Prevailing opinion of a decade ago that Peroxisomes are formed by budding from the ER has had to be abandoned and is now replaced by a model according to which peroxisomes grow by Post-translational import of proteins into existing peroxisomes and new peroxisomes are formed by fission. Complementation analysis of cells from patients with the Zellweger syndrome and related disorders of peroxisome biogenesis has indicated that at least six gene Products are involved in peroxisome assembly. This review concentrates on these new developments.

Characterization of peroxisomes

Microbodies are bounded by a single membrane and generally have a diameter of 0.1 to 1.0 mm and an equilibrium density of 1.21 to 1.25 g/cm^3. They are widespread in nature, and the number of enzymes that can be found i microbodies under various conditions has extended to such an extent that it i no longer meaningful to attempt to tabulate all of them. Table 1 presents a brief summary of microbody nomenclature and characteristic microbody processes. This summary illustrates both the relationships among microbodies and the fact that none of them contains all characteristic microbody processes.

The mammalian peroxisome was originally defined as an organelle containing at least one flavinoxidase producing H_2O_2 and catalase. The term microperoxisome reflects the observation that in many tissues the size of organelles showing a positive cytochemical reaction for catalase is smaller than that of peroxisomes in liver and kidney. Initially, this heterogeneity was the cause of considerable confusion in the localization of enzymes to (micro) peroxisomes after differential centrifugation of homogenates. It is now generally believed that the prefix "micro-" -reflects a difference in size only rather than enzyme content and function, and the term peroxisome will be used indiscriminately in this review. An organelle in germinating castor beans containing the five enzymes of the glyoxylate cycle received the name "glyoxysome". It was then found to have a microbody-like morphology and to contain oxidases, catalase, and a β-oxidation system.

On the basis of these properties, glyoxysomes are considered to be the plant kingdom variant of mammalian peroxisomes. Upon the transition from germinating seeds to seedlings with green leaves, the glyoxysomes develop the capacity for a photorespiration process in which glyoxylate is produced from glycolate and converted by a transaminase to glycine. Initially it was reported that leaf Peroxisomes lacked the β-oxidation system present in glyoxysomes, but more recent data indicate that fatty acid β-oxidation is a general function of higher plant peroxisomes. In fact, microbodies are now regarded as the sole location in plant cells where fatty acid β-oxidation takes place.

Ether-linked phospholipids have been reported in higher plants and their seeds, but only trace quantities are present, generally amounting to less than 1% of the total phospholipids. There have been no reports on the presence in plant glyoxysomes/peroxisomes of acyl-CoA:DHAP-AT, alkyl-DHAP synthase, and NADPH:alkyl-DHAP oxidoreductase, three enzymes required for the synthesis of ether phospholipids in mammalian peroxisomes. Yeast microbodies contain a complete β-oxidation system, several H_2O_2 producing oxidases, and catalase, thus characterizing them as peroxisomes. In addition, the two glyoxylate cycle enzymes, isocitrate lyase and malate synthase, are present when the cells are grown on C2-compounds like acetate and ethanol or on compounds that are degraded to C_2-compounds like alkanes and fatty acids.

The content of other enzymes in yeast microbodies is strongly dependent on the carbon and nitrogen sources used for growth. With the exception of Pullularia pullulans, in swhich choline- and ethanolamine phospholipids have 12% and 32% plasmalogens, respectively, most common yeasts like Saccharomyces cerevisiae and *Hansenula* polymorpha have at best trace amounts of plasmalogens. In addition, the specific peroxisomal DHAP-AT involved in ether lipid biosynthesis in mammalian tissues could not be demonstrated in yeast. Mutants of S. cerevisiae defective in sn-glycerol-3-phosphate AT showed a simultaneous loss of DHAP-AT activity, indicating a common structural gene for both activities. A new function for microbodies in some fungi was recently suggested when the localization of enzymes involved in penicillin biosynthesis was studied in Penicillium chrysogenum.

Immunogold cytochemistry and cell fractionation provided strong evidence that the acyltransferase, which catalyzes the final step in penicillin biosynthesis, is located in organelles with a diameter of 0.2 to 0.8 μm that most probably represent microbodies. The microbody of the aerobic protozoa Tetrahymena pyriformis played an important role in establishing the phylogenetic relationship between peroxisomes and glyoxysomes. Like the yeast peroxisome, it contains the two glyoxylate cycle enzymes, isocitrate lyase and malate synthase, H_2O_2 producing oxidases, catalase, and a complete β-oxidation system. *T.* pyriformis strains do not contain plasmalogens, but are very rich in alkyl-acyl glycerophospholipids. Neither the mechanism nor the localization of the responsible enzymes has been resolved. It should be remembered that the mere presence of ether-linked phospholipids cannot be taken as evidence for the operation of the DHAP pathway for ether lipid synthesis.

In fact, experiments using [2^{-3} H] and [2^{-14}C] glycerol revealed that DHAP did not serve as an intermediate in the synthesis of plasmalogens in anaerobic bacteria and protozoa. In *Trypanosomatides* an organelle resembling peroxisomes in morphology, but lacking peroxisomal marker enzymes like H_2O_2-producing oxidases and catalase, was termed "glycosome" because it harbored seven enzymes of glycolysis. Enzymes of other pathways including ether lipid synthesis and P-oxidation are also present in *Trypanosoma brucei* or closely related organisms.

The β-oxidation system in these and other catalasefree microbodies may use acyl-CoA dehydrogenase rather than acyl-CoA oxidase for the initial desaturation of acyl-CoA. The glycosome of *T. brucei* contains a specific DHAP-AT that does not use glycerol- 3 -phosphate, alkyl/acyl-DHAP oxidoreductase, and acyl-CoA reductase for the formation of long-chain alcohols. This strongly suggests that the glycosome of *T. brucei* shares the DHAP-pathway of ether lipid synthesis with the mammalian peroxisome, although the actual formation of glycero-ether bonds by alkyl-DHAP synthase has not yet been shown. All of the above microbodies are surrounded by a single membrane and do not contain

DNA. An apparently unrelated microbody in *Trichomonatides* received the name "hydrogenosome" because of its capacity to transfer electrons from pyruvatc oxidation to H' under anaerobic conditions.

Hydrogenosomes share with other microbodies a high equilibrium density of 1.24 to 1.26 g/cm^3 but differ in characteristic other properties, i.e. they have been reported to contain DNA and appear to be bounded by a double membrane.

Morphology and induction of peroxisomes

In liver and kidney, peroxisomes appear *As* round or slightly oval structures with a diameter of 0.3 to 1.0 μm. In liver they are conspicuous in the parenchymal cells where they occupy 2.4% of the cellular volume. Hepatic peroxisomes often contain crystalloid cores believed to represent rate oxidase aggregates. This structure is not present in human tissues, which lack rate oxidase. Further support for this notion was provided by protein A-gold immunochemistry, demonstrating that the crystalloid cores in bovine kidney peroxisomes were the exclusive site of urate oxidase localization. By contrast, rat kidney peroxisomes lack urate oxidase and do not contain a crystalloid core.

Marginal plates, another common type of crystalline inclusion in renal peroxisomes, were recently shown to be composed mainly of the B isozyme of L-α-hydroxyacid oxidase. The detection and identification of peroxisomes are facilitated by using 3,3'-diaminobenzidine staining employing the peroxidative activity of catalase at alkaline pH. In most mammalian tissues other than liver and kidney, catalase-positive particles are smaller, with diameters of 0.05 to 0.2 μm, an devoid of a crystalloid core. Peroxisomes are abundant in tissue active in lipid metabolism such as liver, sebaceous glands, and brown fat. I nervous tissue of the rat, they are especially enriched in myelin-producing oligodendrocytes and most abundant in the oligodendrocyte processes adjacent to the growing myelin sheets at day 17 postnatal when myelin formation peaks. A close spatial relationship between peroxisomes and sites of lipid synthesis, such as ER membranes, has frequently been observed. Initially, this close association was misinterpreted as a luminal continuity between both organelles and contributed to the view that peroxisomes were formed by budding from the ER.

This view was revised, in part because of more recent experiments employing serial sectioning electron microscopy. Such experiments demonstrated instead clusters of interconnected peroxisomes that form a peroxisomal reticulum in mouse liver and in regenerating rat liver. These elongated tubular structures provide an explanation for the variation in size and shape of catalase-positive particles in thin sections. Morphological studies using catalase cytochemistry in conjunction with more specific immunocytochemistry have contributed to our understanding of peroxisome biogenesis.

Fahimi and coworkers have used specific antibodies against peroxisomal matrix and membrane proteins to study the gradual reappearance of these proteins after partial hepatectomy. The synthesis of membrane proteins preceded that of matrix proteins, suggesting that peroxisome biogenesis is initiated by the synthesis of new membranes. Additional evidence for this was obtained from peroxisome proliferation in rat liver and yeast. In rats treated with a hypocholesterolemic drug, doublemembrane loops resembling smooth ER were seen in the proximity of some peroxisomes.

Serial sectioning revealed that these loops were continuous with the peroxisomal membranes and that they reacted with antibodies against proteins. These loops may serve as peroxisomal membrane extensions prior to import of peroxisomal matrix proteins. In methylotrophic yeasts, peroxisomes are massively induced upon growth in media containing methanol. In H. *polymorpha*, single peroxisomes mature to one large organelle before up to 18 smaller organelles are split off, which may occupy more than 80% of the cell volume. Similarly, *Candida boidinii* contained two to five small peroxisomes when cultured in glucose media.

A study of their proliferation upon a shift to methanol-containing media indicated that several elongated peroxisomes developed from a preexisting organelle. These incorporated catalase and a 47-kDa membrane protein, but not yet alcohol oxidase, dihydroxy acetone synthase, and a 20-kDa membrane protein. The latter three proteins were imported during further proliferation when up to 30 clustered peroxisomes were formed by division. These findings and those obtained after peroxisome proliferation in liver provide strong morphological evidence that newly formed peroxisomes arise by growth and division of pre-existing peroxisomes. S. *cerevisiae* long seemed to form an exception to the general observation that peroxisomes in yeasts can be induced by changes in growth media.

Peroxisome proliferation in this organism was recently shown to occur upon growth in media containing oleic acid. Glucose-grown cells generally contained one to four small peroxisomes. After the shift to oleic acid media, these developed into up to 14 mature peroxisomes with diameters of up to 0.5 μm. Hypolipidemic drugs, herbicides, and phthalate ester plasticizers induce proliferation of peroxisomes in hepatic parenchymal cells of rodents. Additional morphometric studies reported a doubling in the number and average size of peroxisomes in mouse hepatocytes after four days of clofibrate feeding, resulting in an increase of the cytoplasmic area occupied by peroxisomes from 1.8% to 7.6%. A similar increase, from 1.4% to 5.4%, was observed in rat liver after two weeks of plasticizer feeding. Proliferation can also be induced in vitro in cultured rat hepatocytes to cause the cytoplasmic volume of peroxisomes to increase from less than 1% to 9-12%. No increases in cultures derived from monkey and human were observed.

Since peroxisome proliferation by hypolipidemic drugs has been implicated in liver carcinogenesis, the absence of or limited proliferation in primates may suggest the absence of tumorigenic activity in these species. Reddy and coworkers suggest that the carcinogenicity of peroxisome proliferators is attributable to the increased activity of peroxisomal β-oxidation resulting in overproduction of H_2O_2 and possible oxidative damage to DNA. In this respect, it is interesting to note that a new hypolipidemic drug, 4-(2-[4-(chlorocinnamyl) piperazin-1-yllethyl) benzoic acid, was recently shown to induce marked proliferation of peroxisomes in rat liver without the significant elevation of the β-oxidation capacity that generally accompanies proliferation. The induction of β-oxidation usually reaches a maximum after two weeks of treatment.

The activity versus time profile is nearly identical for the three individual β-oxidation enzymes, i.e. acyl-CoA oxidase, bifunctional enzyme, and thiolase, and is mainly due to an increased rate of synthesis from the enhanced mRNA levels. In comparison to the approximately 20-fold increase in the levels of mRNA for peroxisomal β-oxidation enzymes in rat liver, the increases in kidney, small intestine, and heart were small, varying from 2- to 4-fold, and no increases were observed in nine other extrahepatic tissues. Even in liver, other peroxisomal enzymes such as catalase, urate-, α-hydroxyacid-, and D-amino-acid oxidase exhibit only marginal increases, remain unchanged, or are reduced. Some, but not all, membrane proteins are inducible. The mechanism of action of the proliferators remains largely to be clarified. It has been suggested that a carboxylate function, which is either present or can be formed by oxidation, and which can be activated to a CoA ester, forms a common motif of otherwise structurally nonrelated proliferators.

The detection by affinity chromatography of a peroxisome proliferator binding protein in rat liver, which turned out to be identical with the 72-kDa member of the heat-shock protein family, suggested that these Proliferators could act by a mechanism similar to that of steroid hormones to activate transcription of peroxisomal β-oxidation genes. Following this reasoning, Isseman & Green recently cloned a novel member of the steroid hormone receptor family of ligand-activated transcription factors that is activated by a diverse class of peroxisome proliferators.

The mouse gene coded for a 52-kDa protein and could be expressed in liver, kidney, and heart in a manner that mirrored the tissue-specific effects of peroxisome proliferators. These findings suggest that this receptor directly mediates the effect of such proliferators, although further ligand binding studies are needed. The relationship with the 72-kDa peroxisome proliferator-binding protein from rat liver remains at present unclear.

Selected metabolic pathways

The first peroxisomal function discovered by de Duve and coworkers was its simple respiratory pathway. The pathway is based upon the formation of hydrogen peroxide by a collection of oxidases and the decomposition of the H_2O_2 by catalase. These reactions may be responsible for as much as 20% of oxygen consumption in rat liver. Among the oxidases identified are: urate oxidase, D-aminoacid oxidase, L-α-hydroxyacid oxidase A and B, acyl CoA oxidase, glutaryl-CoA oxidase, polyamine oxidase, pipecolic acid oxidase, oxalate oxidase, trihydroxycholestanoyl CoA oxidase, and pristanoyl-CoA oxidase.

The decomposition of H_2O_2 may occur catalytically by reaction of two molecules of H_2O_2 to form H_2O and O_2 or, alternatively, peroxidatically. Both reactions are catalyzed by catalase. Catalase accepts a range of peroxidation substrates such as ethanol, methanol, nitrites, quinones, and formate. Until recently it was generally considered that ethanol is oxidized predominantly via cytosolic alcohol dehydrogenase. However, as shown by Handler & Thurman, ethanol elimination via catalase may be significant if substrates such as fatty acids are available to generate H_2O_2 inside the peroxisome.

Fatty acid β-oxidation

Cooper & Beevers; were the first to show that glyoxysomes are capable of catalyzing the β-oxidation of fatty acids. Inspired by these studies, Lazarow & de Duve demonstrated the presence of a fatty acid β-oxidation system in rat liver peroxisomes. Subsequent studies have shown that nonmitochondrial β-oxidation is widely distributed in nature, much more so in fact than β-oxidation in mitochondria, which is mainly restricted to the animal kingdom. Nonmitochondrial β-oxidation has been detected in animals, plants, and eukaryotic microorganisms.

Enzymic organisation: Since acyl-CoAs rather than free fatty acids are the true substrates for β-oxidation, conversion to COA-esters is the first obligatory step. This is brought about by a variety of acyl-CoA synthetases, which differ with regard to their substrate specificity and subcellular localization. Long-chain fatty acid activation can occur via a long-chain acyl-CoA synthetases, which is located at the cytosolic face of the peroxisomal membrane. An identical enzyme is present on the outer mitochondrial membrane and the membrane of the ER. Recently, evidence has been brought forward suggesting that liver peroxisomes contain a second acyl-CoA synthetase, which preferentially activates very long-chain fatty acids (VLCFA) such as tetracosanoic and hexacosanoic acid.

A similar activity was detected in the ER but not in mitochondria. There is disagreement with regard to the topological localization of the VLCFA activating enzyme in the peroxisomal membrane. Whereas Lazo et al reported that the active site faces the peroxisomal interior, Lageweg *et al* found a cytosolic orientation. Studies on the human genetic disorder X-ALD indicated a deficiency of the peroxisomal VLCFA-CoA synthetase but normal activity of the ER enzyme. This suggests an obligatory coupling between the activation of VLCFA on the peroxisomal membrane and the subsequent β-oxidation of VLCFA within this organelle.

Elegant studies by Numa and coworkers have shown a similar type of compartmentation of the metabolism of oleic acid in the yeast *Candida lipolytica*. However, it is not a general principle of fatty acid degradation in peroxisomes that activation and subsequent β-oxidation are coupled processes. Activation of di- and trihydroxycholestanoic acid, dicarboxylic acids, and prostaglandins occurs at the ER. Following activation, the acyl-CoAs must be carried across the peroxisomal membrane via a mechanism that is as yet not well understood. Once inside the peroxisome, CoA-esters are subjected to oxidation by acyl-CoA oxidases. Like the mitochondrial acyl-CoA dehydrogenases, these enzymes contain FAD as a prosthetic group. Reoxidation of enzyme-bound $FADH_2$, however, occurs via direct interaction with molecular oxygen, yielding H_2O_2. Acyl-CoA oxidases have been purified from a number of sources view including rat liver, cucumber seeds, and different yeast species, and possess very similar subunit sizes. In rat liver, two mRNAs for acyl-CoA oxidase were identified, which are produced by alternative splicing of the primary transcript of a single acyl-CoA oxidase gene.

Comparison of the acyl-CoA oxidase sequences from rat and different yeast specie shows significant homology. Recently, evidence was reported to the existence of two acyl-CoA oxidases in rat liver peroxisomes, one of which is clofibrate- inducible and the other not. Whether these acyl-CoA oxidases are the products of the two mRNAs identified by Miyazawa et al is not yet known. Apart from these acyl-CoA oxidases, there is a separate oxidase catalyzing the first step in trihydroxycholestanoyl-CoA oxidation. The latter enzyme is expressed in liver only and is not induced by clofibrate. The next two reactions in peroxisomal β-oxidation are catalyzed by multifunctional proteins, which, apart from enoyl-CoA hydratase and L-3-hydroxyacyl-CoA dehydrogenase activities, may also contain other enzyme activities.

A trifunctional β-oxidation protein harboring 3-hydroxyacyl-CoA epimerase as a third enzyme activity was found in *Candida tropicalis*, C. lipolYtica, and *Neurospora* crassa, whereas a bifunctional protein was found in S. *cerevisiae*. The rat liver enzyme, long known as a bifunctional protein, is in fact a trifunctional protein having Δ^3-*cis* Δ^2-*trans*-enoyl-CoA isomerase as third inherent activity. Molecular cloning of the cDNA for rat liver trifunctional protein has revealed strong sequence similarity with both mitochondrial hydratase and mitochondrial 3-hydroxyacyl-CoA dehydrogenase, suggesting that the gene has originally been produced by gene fusion. Finally, the last reaction in peroxisomal β-oxidation is catalyzed by 3-oxoacyl-CoA thiolase. Peroxisomal thiolases have been purified from a number of sources including rat liver, cucumber seed glyoxysomes, and several yeast species. In contrast to the mitochondrial and cytosolic thiolases, which are tetramers, peroxisomal thiolases are dimeric with a subunit size of 40-50 kDa . In rat liver there are two peroxisomal thiolase species differing with respect to molecular weight and inducibility by clofibrate.

These two thiolases are encoded for by different genes, which show strong sequence homology. Apart from the enzymes directly involved in β-oxidation, peroxisomes also contain the auxiliary enzymes required for the degradation of unsaturated fatty acids, i.e. 2,4-dienoyl-CoA reductase and Δ^3-*cis*-Δ^2 -*trans*-enoyl-CoA isomerase. Recently, Schulz and coworkers identified a novel enzyme in peroxisomes, D-3-hydroxyacyl-CoA dehydratase, which may function in the β-oxidation of unsaturated fatty acids and/or D-hydroxy fatty acids. Peroxisomes also possess medium-chain carnitine acyltransferase activity and carnitine acetyltransferase activity.

Physiological function of the peroxisomal b-oxidation system: The peroxisomal β-oxidation system is not just a functional duplicate of the mitochondrial system. It is now clear that peroxisomes are involved in the degradation of a distinct set of compounds. With regard to long-chain saturated fatty acids, it is generally agreed that at least in rat liver, oxidation takes place predominantly in mitochondria, with estimates varying from 70% to more than 95%.

The initial steps involved in the P-oxidation of VLCFA proceed in peroxisomes rather than in mitochondria. This might well be due to the fact that mitochondria lack the capability to activate VLCFA. Recently, it has been argued that although oxidation of VLCFA may be peroxisomal in rat liver and human skin fibroblasts, it may be mitochondrial in brain.

Like mitochondria, peroxisomes possess 2,4-dienoyl-CoA reductase activity and Δ^3-*cis*Δ^2 -trans-enoyl-CoA isomerase activity, thus enabling them to handle the double bonds in unsaturated fatty acids. Studies on the β-oxidation of fatty acids with chain lengths of 20 and 22 carbon atoms and possessing from zero to six double bonds have shown that in general the peroxisomal β-oxidation system shows a preference for the more polyunsaturated fatty acids. No quantitative analysis has been made in intact cells to evaluate the contribution of peroxisomes and mitochondria to overall *Poxiclation* of these polyunsaturated fatty acids.

However, recent studies using peroxisome-deficient cell lines indicated that the oxidation of erucic acid, adrenic acid, arachidonic acid, and tetracosatetraenoic acid is initiated in peroxisomes and not in mitochondria. Also, the β-oxidation of branched-chain fatty acids, such as pristanic acid, appears to occur in peroxisomes. In addition, peroxisomes have been implicated in the β-oxidation of dicarboxylic acids, prostaglandins, and xenobiotics with an acyl side-chain. It is as yet uncertain, however, whether or not peroxisomes are obligatory in the β-oxidation of these compounds.

Glyoxylate metabolism

The glyoxylate cycle, discovered by Kornberg & Krebs, provides a mechanism for the net conversion of acetyl units to carbohydrate in microorganisms and germinating fat-bearing seeds. This conversion is brought about by malate synthase and isocitrate lyase acting together with enzymes of the citric acid cycle. The glyoxylate cycle is generally considered to be absent from animal tissues. Goodman and coworkers, however, have reported that isocitrate lyase and malate synthase are present in several vertebrate tissues, including adipose tissue of the hibernating bear and rat liver.

The activities found by these authors are surprisingly high. While a role for the glyoxylate cycle could be envisaged in the former case, this pathway is unlikely to be of quantitative significance in rat liver in view of the fact that gluconeogenesis from C_3 compounds in liver and kidney is essential for glucose homeostasis in mammals. In mammals glyoxylate can be metabolized by conversion to glycine in a reaction catalyzed by AGT, an enzyme detected only in liver and to a lesser extent in kidney. In this way glyoxylate enters the gluconeogenic pathway. This is an essential metabolic pathway in humans, as indicated by the fact that an inherited deficiency of the enzyme leads to primary hyperoxaluria type 1, a serious disease caused by massive production of oxalate from glyoxylate.

The subcellular distribution of AGT varies with species and tissues. The enzyme in liver is exclusively present in peroxisomes in humans, baboons, rabbits, and guinea pigs. A peroxisomal location is also found in guinea pig kidney, but not in rabbit kidney, where the enzyme is found nearly exclusively in mitochondria, as is the case in dog and cat liver. Despite the fact that the localization of the enzyme in dogs differs from that in humans, primary hyperoxaluria type I has also been encountered in dogs. AGT in marmoset and rat and mouse liver is present both in peroxisomes and in mitochondria.

The peroxisomal and mitochondrial forms have identical physicochemical and enzymological characteristics and are immunologically indistinguishable.

Indeed, recent molecular genetic studies have shown that the two forms are products of a single gene with the mRNAs being transcribed from different initiation sites. In rats, the peroxisomal enzyme is induced by feeding with clofibrate or plasticizer, whereas the mitochondrial form is induced by glucagon.

Cholesterol and dolichol metabolism

Mammalian peroxisomes have recently been implicated as a site of cholesterol biosynthesis. Immunoelectronmicroscopy and subcellular fractionation studies demonstrated that substantial amounts of HMG-CoA reductase could be detected in peroxisomes. HMG-CoA reductase in the ER is considered to be the key regulatory enzyme of isoprenoid biosynthesis. The molecular masses of the peroxisomal and the ER enzyme proteins are apparently identical, even though the ER enzyme is a glycosylated transmembrane protein whereas the peroxisomal enzyme is a soluble protein found in the matrix space of the organelle.

Highly purified peroxisomes contain a thiolase activity that is able to catalyze the initial step in cholesterol synthesis, i.e. the condensation of acetyl-CoA units to acetoacetyl- CoA, and are able to convert mevalonic acid to cholesterol in the presence of a cytosolic fraction in vitro. The presence of individual enzymes of cholesterol biosynthesis in rat liver peroxisomes was also reported by Appelkvist et al. The physiological and quantitative significance of this putative peroxisomal pathway for cholesterol synthesis is not yet clear. Peroxisomes in liver play an essential role in the biosynthesis of bile acids. The conversion of the intermediate 3α,7α,12α-THCA to cholic acid is a peroxisomal process involving oxidative cleavage of the side-chain with release of propionyl-CoA.

It is of interest that other proteins thought to be involved in cholesterol utilization are also present in peroxisomes. Especially the subcellular localization of sterol carrier protein 2 has been studied in detail. SCP-2 is a nonenzymic protein of 13.5 kDa involved in the conversion of lanosterol to cholesterol, the esterification of cholesterol, and the synthesis of bile acids. It is also involved in the intracellular transfer of cholesterol, which is required in pregnenolone production in adrenal and ovarian tissues. Immunoelectronmicroscopic studies in rat liver demonstrated that SCP-2 is present inside mitochondria and associated with the ER and the cytosol, but the largest concentration of SCP-2 is inside the peroxisomes.

The dependence of SCP-2 on the peroxisomal compartment is also illustrated by the greatly reduced amount of protein in liver homogenates of Zellweger patients and in CHO mutants deficient in peroxisomes. Cloning and expression of cDNAs encoding rat liver and human liver SCP-2 indicated that the protein is synthesized in both species as a 143-amino-acid long pre-SCP-2, including a 20-residue N-terminal leader sequence. Molecular cloning also detected much larger cDNAs derived from the same gene that encoded a 59-kDa protein containing the sequence of pre-SCP-2 as its C-terminal part. This protein has consistently been found in peroxisomes using antibodies against SCP-2 . As pointed out by Mori et al, the presence of this protein in peroxisomes makes it unlikely that it is processed to SCP-2, and its function needs to be studied further.

Notably, apart from containing the SCP-2 sequence, the 59-kDa protein has significant sequence homology with mitochondrial and peroxisomal 3-oxoacyl-CoA thiolases. The implications of these observations are presently not clear, but suggest that the 59-kDa protein may somehow be related to β-oxidation. In this respect, it is worth noting that a peroxisomal nsLTP from C. *tropicalis*, showing 33% homology with nsLTP from rat liver, was exclusively expressed in cells grown on oleate, also suggesting a role in the β-oxidation of fatty acids. The initial steps in the biosynthetic pathway leading to cholesterol are also involved in dolichol biosynthesis. Based on in vitro and in vivo experiments.

Ether lipid synthesis

Localization and topology: The pathway of glycero-ether bond formation was elucidated two decades ago, and starts with the acylation of DHAP. A recent claim that this replacement could also take place in the nonphosphorylated acyl- dihydroxyacetone appeared to be based on erroneous product identification. The initial confusion about the subcellular localization of DHAP-AT, i.e. in the crude mitochondrial fraction of liver and in the microsomal fraction of several extrahepatic organs, was nicely resolved when Hajra and coworkers showed that the enzyme was in fact mainly localized in peroxisomes.

The presence of the DHAP-pathway enzymes in peroxisomes suggested an important role of this organelle in glycero-ether lipid synthesis. This importance only became fully appreciated after we demonstrated that tissues from Zellweger patients, known to lack peroxisomes, were severely deficient in ether-linked phosphoglycerides, and that fibroblasts from such patients were impaired in plasmalogen biosynthesis. This latter condition appeared to be caused by a deficiency in both DHAP-AT and alkyl-DHAP synthase.

The essential role of peroxisomes in ether lipid synthesis was also established in CHO mutants defective in peroxisome biogenesis. In all cases the deficient ether lipid synthesis could be corrected by supplementation of the cells with alkylglycerol, which convincingly demonstrated that the deficiency in ether lipids in these mutant cells was solely caused by impaired formation of the glycero-ether bond in peroxisomes, and indicated that ether lipid synthesis cannot take place anywhere else in the cell to any appreciable extent. Acyl-CoA can be provided by an acyl-CoA ligase, the active site of which faces the cytoplasm. By contrast, DHAP-AT and alkyl-DHAP synthase are not sensitive to proteolysis unless detergent is used to disrupt the membrane, suggesting a matrix-oriented localization of the active sites of these enzymes. For DHAP-AT this conclusion was recently supported by experiments with selectively permeabilized fibroblasts. Under these conditions DHAP-AT showed latency that was not observed in sonicated cells.

This latency could be overcome by addition of ATP, suggesting that ATP is required for the transport of at least one of the substrates of DHAP-AT over the peroxisomal membrane to reach the active site of this enzyme at the inner aspect of the membrane. It has been suggested that a fatty alcohol-producing acyl-CoA reductase is present in rat brain peroxisomes, but it is generally believed that this enzyme is located in microsomes.

The fact that this membrane associated enzyme retains near normal activity in ZS fibroblasts would be in line with an extra peroxisomal localization. Once alkyl-DHAP has been formed, presumably at the inner aspect of the peroxisomal membrane, it has to be translocated for reduction by acyl/alkyl-DHAP reductase located at the cytoplasmic side of either peroxisomes or ER. Acyl-DHAP and alkyl-DHAP generated in isolated liver peroxisomes transverse the membrane within minutes to become available at the outer surface for extraction with bovine serum albumin. The next step in complex ether lipid synthesis concerns the acylation of the *sn*-2-hydroxyl. This and further enzymes in the pathway are not found in peroxisomes, and the process is completed in the ER. It is not known whether alkyl-DHAP, alkyl-glycerol- 3 -phosphate, or both will leave the peroxisome, nor whether the transport is protein-mediated or involves monomer diffusion. It is clear, however, that the function of peroxisomes is to provide ether-linked intermediates for the final steps of ether lipid biosynthesis in the ER. Peroxisome proliferation in liver by clofibrate or plasticizer results in a 2-to 3-fold increase in total and specific activity of DHAP-AT and a less than 2-fold increase in acyl/ alkyl-DHAP reductase.

However, in purified peroxisomes the specific activity of DHAP-AT exhibited no significant increase under conditions causing an at least 10-fold increase in fatty acid β-oxidation specific activity. In homogenates the specific activity of alkyl-DHAP synthase was hardly affected by proliferation, and a slight decrease in purified peroxisomes was observed. In spite of the increase in total activity of the enzymes per liver cell, the plasmalogen content did not change appreciably upon proliferation.

Enzymology: Neither DHAP-AT nor alkyl-DHAP synthase have been obtained in homogeneous form. A partially purified preparation of DHAP-AT from guinea pig liver peroxisomes was devoid of glycerol - 3 -phosphate acyltransferase activity. The biosynthetic reaction is reversible as shown by the formation of palmitoyl-[^{3}H]CoA from palmitoyl-DHAP and [^{3}H]CoA. In addition, the enzyme catalyzes the CoA-dependent exchange of the acyl group between acyl-DHAP and DHAP. The mechanism of the unique reaction catalyzed by alkyl-DHAP synthase has received considerable attention. The use of partially purified enzyme preparations, devoid of contaminating acyl hydrolases, has been particularly rewarding. Characteristic features in the formation of the glycero-ether bond are the loss of the pro-R hydrogen at the esterified C-atom in acyl-DHAP and the stoichiometric replacement of the acyl group by a fatty alcohol donating the oxygen for the ether linkage.

These features predicted an unusual ester cleavage between the C-atom of DHAP and oxygen rather than the usual cleavage between oxygen and the carbonyl carbon of the fatty acid. This exceptional cleavage mechanism was confirmed experimentally using palmitoyl--[^{18}O]DHAP. Acyl/alkyl-DHAP reductase has recently been obtained in homogenous form as a 60-kDa protein after 370-fold purification from guinea pig liver peroxisomes.

The enzyme reduces acyl-DHAP and alkyl-DHAP with nearly equal K_M and V_{max} values, and exhibits a large preference for NADPH over NADH with K_M values of 20 μM and 1.7 mM, respectively.

Peroxisomal disorders

In humans the vital importance of peroxisomes received particular emphasis by the recognition of a number of inherited diseases involving either a defect in the biogenesis of peroxisomes or a deficiency of one (or more) peroxisomal enzymes. So far 12 different peroxisomal disorders have been described, most of them lethal diseases. In 1973 Goldfischer and coworkers showed that morphologically distinguishable peroxisomes are absent in liver and renal tubule cells of patients suffering from the cerebro-hepato-renal (Zellweger) syndrome. This finding was later confirmed by other investigators, who also showed that the deficiency of peroxisomes is manifested in all tissues and cells studied so far, including cultured skin fibroblasts. The absence of peroxisomes in the Zellweger syndrome is accompanied by the following biochemical abnormalities.

— The biosynthesis of ether phospholipids is impaired, resulting in a deficiency of plasmalogens in all tissues.

— The β-oxidation of fatty acids is impaired due to a deficiency of all peroxisomal b-oxidation enzyme proteins. This impairment results both in an accumulation of very long-chain fatty acids in tissues and blood and of intermediates in the biosynthesis of bile acids in blood. Schram *et al* have shown that in Zellweger fibroblasts the peroxisomal b-oxidation enzyme proteins are synthesized normally but become rapidly degraded thereafter.

— There is a defective degradation of pipecolic: acid due to a deficiency of pipecolic acid oxidase.

— The degradation of phytanic acid is impaired due to a deficiency at the level of phytanic acid a-oxidation. Fifthly, several peroxisomal matrix enzymes like catalase, D-aminoacid oxidase, L-a-hydroxyacid oxidase, and AGT are localized. In the cytosol instead of being particle-bound. Recent studies by several groups have shown that some of the peroxisomal integral membrane proteins are present in fibroblasts from Zellweger patients in unusual, largely empty membrane structures called peroxisomal "ghosts".

Classification

Peroxisomes can tentatively be classified into three groups depending on the degree to which peroxisomal unctions are deficient. The first category consists of disorders like the Zellweger syndrome with a generalized impairment of peroxisomal functions, and also comprises two clinically somewhat milder disorders, the neonatal type of adrenoleukodystrophy (NALD) and the infantile type of Refsum disease (IRD). There may be heterogeneity with respect to the degree of peroxisome deficiency. In cultured skin fibroblasts from some Zellweger and IRD patients and in intestinal epithelia from several Zeilweger and NALD patients, the presence of a limited number of peroxisomes has been demonstrated by ultrastructural enzyme cytochemistry.

A remarkable genetic heterogeneity was found within this category by complementation after somatic cell fusion of fibroblasts. Complementation was assessed by measuring DHAP-AT activity, particle-bound catalase, de novo plasmalogen biosynthesis, phytanic acid oxidase activity, and peroxisomal β-oxidation activity. The results indicate that at least six different gene products play a role in peroxisome biogenesis. A second category comprises two diseases, the rhizomelic type of chondrodysplasia punctata (RCDP) and Zellweger-like syndrome.

A limited number of peroxisomal functions is impaired in these disorders. In RCDP the de novo plasmalogen biosynthesis is deficient, phytanic acid oxidation is impaired, and the peroxisomal thiolase is present only as a 44-kDa precursor protein. Most of this precursor thiolase is extraperoxisomal. However, no accumulation of very long-chain fatty acids or intermediates of bile acid biosynthesis is found, indicating that the peroxisomal, β-oxidation capacity is sufficient to prevent accumulation of substrates.

Complementation studies indicate that RCDP is distinct from Zellweger syndrome, NALD, and IRD. In Zellweger-like syndrome, a very rare disorder, peroxisomes appear to be present, but the peroxisomal β–oxidation enzyme proteins were found to be deficient upon immunoblotting, there was deficiency of DHAP-AT, and the content of plasmalogens in tissues and cultured cells was low.

The third category includes a larger group of genetic disorders in which only one peroxisomal enzyme is deficient and in which normal peroxisomes can be found in tissues and cultured cells in most patients. The most commonly encountered disorder in this category is X-linked adrenoleukodystrophy (X-ALD) and its phenotypic variants, in which very long-chain fatty acids (VLCFA) accumulate in tissues, plasma, and cultured cells. A key observation was that in VALD fibroblasts, the oxidation of lignoceroyl-CoA was unimpaired, in contrast to the oxidation of free lignoceric acid. Subsequent studies confirmed that the biochemical defect in X-ALD is at the level of the peroxisomal VLCFA-CoA synthetase.

Patients have recently been described with either a deficiency of acyl-CoA oxidase, the bifunctional protein, or peroxisomal thiolase. In some patients the deficiency was established by the absence of immunoreactive protein in immunoblots. In others, immunoreactive protein is present, and the mutation presumably affects the catalytic activity of the enzyme. In patients with a deficiency of the bifunctional protein or of peroxisomal thiolase, not only VLCFA, but also the bile acid intermediates DHCA and THCA accumulate. In contrast, bile acid intermediates do not accumulate in patients with an acyl-CoA oxidase deficiency, which is in accordance with the recent finding of separate oxidases for THCA-CoA and VLCFA-CoA. Patients with an accumulation of THCA and DHCA but not of VLCFA have also been described, presumably reflecting a defect in THCA-CoA oxidase or THCA-CoA synthetase.

Hyperoxaluria type I can also be classified as a peroxisomal disorder of the third category. It has recently been shown that the defect in this disorder is a deficiency of the enzyme AGT, which is a peroxisomal enzyme in humans. Interestingly, some patients have a trafficking defect resulting in diversion of AGT to the mitochondria.

Finally, acatalasemia is a condition in which the peroxisomal enzyme catalase is deficient. Individuals with a deficiency of catalase usually do not show severe clinical abnormalities.

Assembly of peroxisomes

Introductory remarks and genes for peroxisomal proteins

It is now well-established that peroxisomes, like mitochondria and chloroplasts, arise by growth and division of pre-existing peroxisomes, rather than by budding from the ER. They may transiently or permanently be interconnected to form a "peroxisomal reticulum" containing elements resembling smooth ER. Unlike mitochondria and chloroplasts, peroxisomes contain no DNA, so that all peroxisomal proteins must be coded for by nuclear genes during the past few years, the genes for a large number of peroxisomal proteins, including mammalian peroxisomal β-oxidation enzymes, have been cloned and characterized. No features common to all of these genes are as yet apparent, except perhaps for the presence of GC boxes in the promoter region. In the rat a single acyl-CoA oxidase gene is present, which gives rise to two different mRNAs by differential splicing of the primary transcript.

These two mRNAs may correspond to the two fatty acyl-CoA oxidases identified in rat liver by Schepers *et al,* one of which is inducible by clofibrate in rodent liver. Only one gene for enoyl-CoA hydratase:3-hydroxyacyl-CoA dehydrogenase has been identified in the rat. This gene is inducible, and there is no evidence for the presence of more than one mRNA or more than one protein. Finally, two genes for peroxisomal 3-oxoacyl-CoA thiolase are present in the rat, one of which is inducible by clofibrate.

Thus, entirely different mechanisms are responsible

for the induction by clofibrate of individual β-oxidation enzymes, e.g. (a) enhanced synthesis of a specific mRNA formed by alternative splicing of the transcript of a single gene (acyl-CoA oxidase), or (b) enhanced transcription of a separate gene (3-oxoacyl-CoA thiolase), or (c) enhanced transcription of a single gene (enoylCoA hydratase: 3-hydroxyacyl-CoA dehydrogenase). Only a single gene for peroxisomal 3-oxoacyl-CoA thiolase appears to be present in humans, and only one fatty acid acyl-CoA oxidase, with properties similar to those of the noninducible isoform in rat liver. This may be related to the fact that compounds like clofibrate do not induce peroxisomal β-oxidation enzymes in primates.

In this respect it will be of interest to see if the lack of response in primates correlates with an absence of the 70-kDa peroxisome proliferator-binding protein in rat liver described by Alvarez *et al* or of the 52.4-kDa peroxisomal proliferator-activated receptor in mouse liver recently identified by Isseman & Green.

Peroxisomal proteins

General characteristics of biosynthesis and import All peroxisomal proteins investigated so far, including soluble matrix proteins core proteins, and integral membrane proteins, are synthesized on free ribosomes and are imported posttranslationally into peroxisomes with half-lives in the cytosol of 1-15 min. Almost all are synthesized in their final size. Exceptions are discussed below.

The criteria required for demonstrating import of peroxisomal proteins, both in vivo and in vitro, have been summarized recently. Convincing evidence has been presented for the import by isolated peroxisomes of a number of peroxisomal proteins synthesized in vitro. Import is time-dependent, occurs at 25°C or 37°C but not at 0°C, and requires ATP. The question of whether protein import into peroxisomes, like that into mitochondria, requires a membrane potential is still unresolved. There is some evidence for a requirement for cytosolic factors (unfoldases?) for protein import into peroxisomes.

A topogenic singal for protein import into peroxisomes

They subsequently showed that the targeting signal for directing luciferase to the peroxisomes resides in the C-terminus of the protein, and identified the C-terminal tripeptide -Ser-Lys-Leu (SKL) as the peroxisome targeting signal not only in luciferase but also in other peroxisomal proteins. Gene fusion experiments indicated that reporter proteins with this sequence fused to their C-terminus are directed to the peroxisomes. Certain conservative substitutions are permitted in the first two amino acids (but not in the third) without loss of targeting activity. The SKL motif is present at the C-terminus of a number of peroxisomal proteins from several different organisms. Antibodies raised to a synthetic peptide containing SKL at the C-terminus react with 15 to 20 protein bands in an immunoblot of rat-liver peroxisomal proteins. It has now been demonstrated that the SKL motif is required and is usually sufficient to direct a large number of peroxisomal proteins to the peroxisomes.

Furthermore, this targeting signal has been conserved during evolution and functions in insects, mammals, plants, and different yeasts. In some proteins deletion of certain sequences leads to diminished import even though the C-terminal SKL is intact. This may be due to a change in the conformation of the protein, leading to diminished exposure of the SKL motif. In some peroxisomal proteins the SKL motif (or a variant of it) is not present at the C-terminus but at an internal position in the protein. In human catalase, for instance, which contains 526 amino acids, the sequence -Ser-His-Leu is present at residues 516-518. Gould *et al* have shown that the C-terminal 27 amino acids of catalase acted as a peroxisome targeting signal when fused to a reporter protein, whereas two longer peptides containing the C-terminal 27 amino acids did not. The fact that addition of one or two amino acids to the C-terminus abolishes peroxisome targeting activity suggests that an internal SKL should not function as a peroxisome targeting signal.

Evidence for other peroxisome targeting signals

The fact that the SKL motif or one of its variants is not present in all peroxisomal protein's Indicates that such proteins must use some other peroxisome targeting signal. In a recent paper, Subramani's group report that the peroxisome targeting signal in the precursor of peroxisomal 3-oxoacyl-CoA thiolase is located in the N-terminal peptide that is removed when the precursor is converted to the mature form. Removal of the N-terminal peptide takes place, at least in part, after import. The targeting of glycolytic enzymes to the glycosome, a microbody-like organelle in the Trypanosomatidae, appears to involve different signals from those utilized for peroxisome biogenesis in other eukaryotes. However, proteins carrying the peroxisomal targeting SKL-motif have recently been identified in glycosomes.

Differences among species in the targeting of alanine

As indicated in the section on *Glylate Metabolism,* AGT is found exclusively in peroxisomes in humans and some other species, nearly exclusively in mitochondria in the dog and cat, and in both organelles in rodents. Recent molecular genetic studies have led to some clarification of the molecular genetic mechanisms underlying these differences. Two mRNAs for AGT are found in rat liver. These mRNAs arise from a single gene. Transcription from two initiation sites, 66 nucleotides apart in exon 1, generates the two mRNAs; the larger one contains a

5'-terminal sequence coding for a 22-amino-acid N-terminal extension, which functions as a cleavable mitochondrial targeting signal. Generation of the larger mRNA is under control of glucagon. Both the mRNAs contain the nucleotide sequence coding for an internal SKL sequence, which may act as a peroxisome targeting signal.

Apparently the mitochondrial targeting signal is dominant in the protein derived from the larger mRNA. Human liver contains a single mRNA for AGT in which the untranslated sequence upstream of the first AUG codon shows considerable similarity to the sequence encoding the N-terminal 22-amino-acid peptide of the rat mitochondrial enzyme. In fact, a single base substitution has led to the loss in humans of the AUG initiation codon which, in the rat, allows translation of a protein with the N-terminal extension. Danpure and coworkers have found that in certain patients with primary hyperoxaluria type I, the deficiency of AGT activity is accompanied by an aberrant localization of the enzyme, the residual activity being present in the mitochondria rather than in the peroxisomes. It is not clear at present whether the aberrant routing is due to loss of the peroxisomal targeting signal, appearance of a putative mitochondrial targeting signal, or both.

All such patients investigated so far have at least one allele with point mutations giving rise to amino acid substitutions at positions 11, 170, and 340. The aberrant protein in mitochondria from the patients has the same M_r as that of AGT from control subjects, whereas in rat liver there is a difference in size of about 2 kDa between the precursor and mature form of mitochondrial AGT. Thus the mitochondrial localization of AGT in the patients is apparently not accompanied by detectable cleavage of a targeting signal.

Peroxisomal proteins synthesized as larger precursors: Whereas most peroxisomal proteins are synthesized in their final size there are some exceptions. Peroxisomal 3-oxoacyl-CoA thiolase in mammals is synthesized as a 44-kDa precursor that is converted to a 41-kDa mature enzyme and mammalian nsLTP is synthesized as a 14.5-kDa protein whereas the mature protein is about 1.5-kDa smaller.

In both cases processing does not occur in mutant cells defective in peroxisome biogenesis. Comparison of the amino acid sequence of human and rat peroxisomal 3-oxoacyl-CoA thiolases and nsLTP indicates that there is a striking homology in the region of the peptide bond cleaved during conversion of the precursors to the mature proteins table 2. For instance, proline is present at the P-2 position in all cases. This homology suggests that a single protease, located in peroxisomes, is responsible for the processing of both proteins.

Table 2

Amino acid sequence in the region of the peptide bond cleaved during the processing of the precursor to the nature forms of rat 3-oxoacyl-CoA thiolases A and B, human 3-oxoacyl-CoA thiolase, and human and rat nonspecific lipid transport protein (nsLTP)

Protein	*Amino acid sequence*[a]
Rat thiolase A	-Gln-Ala-Ala-Pro-Cys-!Ser-Ala-Thr-
Rat thiolase B	-Gln-Ala-Ala-Pro-Cys-*Ser-Ala-Gly-
Human thiolase	-Gin-Ala-Ala-Pro-Cys.!Leu-Ser-Gly-
Rat nsLTP	-Ser-Ala-Ala-Pro-Thr-!Ser-Ser-Ala-
Human nsLTP	-Glu-Ala-Val-Pro-Thr!Ser-Ser-Ala-

As indicated in the subsection evidence for other peroxisome targeting signals, the N-terminal peptide of 3-oxoacyl-CoA thiolase appears to contain the singal for directing this protein to peroxisomes. A third peroxisomal protein in mammals that behaves aberrantly is acylCoA oxidase. This enzyme is synthesized as a 72-kDa precursor, which is slowly converted to two subunits of 52- and 20-kDa, both of which remain associated with the enzyme. Again, cleavage of the precursor is impaired in mutant cells lacking peroxisomes. Finally, the isoform of malate dehydrogenase found in glyoxysomes in germinating cotyledons of pumpkin and watermelon is synthesized as a larger precursor with a cleavable presequence

Receptor and accessory proteins required for import: Important of proteins into peroxisomes requires at least two receptors, one recognizing proteins into peroxisomes requires at least proteins with a C-terminal SKL motif and the other recognizing the targeting signal in the N-terminal region of the precursor of 3-oxoacyl-CoA thiolase and perhaps nsLTP. There is as yet no information available about the nature and properties of those putative receptors.

Furthermore, translocation of proteins across the peroxisomal membranes, like that across the mitochondrial membranes can be expected to require the participation of accessory proteins such as unfoldases, channel-forming proteins, and chaperonins. In patients with the disease RCDP, the amount of 3-oxoacyl-CoA thiolase protein in the peroxisome is reduced and processing of the precursor is impaired.

The primary lesion in this disease could be a mutation in a gene coding for a receptor required for the import into peroxisomes of a limited number of proteins, including 3-oxoacyl-CoA thiolase and a putative prothiolase protease.

Peroxisome biogenesis

Transformation of glyoxysomes into leaf peroxisomes in higher plants: During germination of fat-storing seeds, biogenesis of glyoxysomes the glyoxysomes are converted into peroxisomes characteristic of green leaves. The following changes occur. Glyoxysome-specific enzymes, like malate synthase, are broken down in the transition-state organelles present in greening cotyledons. Concomitantly, light induces an increase in the level of mRNA for leaf peroxisome-specific enzymes like glycolate oxidase and in the level of these enzymes in the transitionstate organelles.

Malate synthase synthesized in vitro and imported into isolated microbodies is stable in glyoxysomes and leaf peroxisomes. Thus the enzyme system involved in the degradation of glyoxysomal enzymes is only transiently expressed.

Mutant mammalian cells defective in peroxisome biogensis: Mutant cell lines defective in peroxisome biogenesis are available from patients with the Zellweger syndrome and related disorders and have in addition been isolated from mutagenized CHO cells. Fusion of complementary cell lines leads to correction of peroxisome biogenesis in the heterokaryons as assessed by restoration of peroxisomal functions like DHAP-AT activity, plasmalogen biosynthesis, oxidation of VLCFA, or processing of the precursor of peroxisomal 3-oxoacyl-CoA thiolase to the mature enzyme.

Complementation analysis following somatic cell fusion indicates that the human cell lines can be divided into at least six complementation groups. Thus peroxisome biogenesis requires the participation of at least six gene products. Peroxisome biogenesis in the heterokaryons can also be assessed by measuring the rate of incorporation of catalase, originally present in the cytosol in the parental cell lines, into newly assembled peroxisomes. After fusion of some combinations of human cell lines, the assembly of peroxisomes measured by this criterion was found to be very rapid and insensitive to cycloheximide, so that the components required for peroxisome assembly must have been present in the parental cell lines. Uptake of catalase into peroxisomes in the heterokaryons is inhibited by aminotriazole, presumably because this irreversible inhibitor of catalase prevents or retards unfolding of the enzyme.

Evidence for the presence of peroxisomal ghosts in mutant mammalian cells defective in peroxisome biogenesis: It has been concluded that the lesion in diseases of peroxisome biogenesis must involve the import of peroxisomal proteins rather than the assembly of peroxisomal membranes. This conclusion is based in part on the fact that peroxisomal integral membrane proteins can be detected by means of immunoblotting or immunocytochernistry in liver and fibroblasts from patients with diseases of peroxisome assembly.

The proteins are associated with membranes in large, vesicular structures sometimes devoid of content, which have been referred to as peroxisomal ghosts. However, most of these structures also contain lysosomal enzymes and must therefore represent autophagocytosed material in secondary lysosomes. Nevertheless, in the somatic cell fusion experiments described in the previous section, the assembly of peroxisomes presumably requires the presence of some kind of scaffolding on which assembly of peroxisomal membrances can occur.

Furthermore, it is possible that lack of complementation after fusion of two cell lines results from a lack of a scaffolding. Evidence for the absence of such a scaffolding in a CHO mutant has been presented.

Identification and characterization of genes and gene products required for peroxisome biogenesis: Fujiki and coworkers have recently succeeded in restoring peroxisome biogenesis in a CHO mutant by transfecting the cells with rat-liver cDNA, and have identified the factor responsible for correction as a 35-kDa peroxisomal membrane protein. The function of the protein, which the authors refer to as peroxisome assembly factor PAF-I, is as yet unknown. An analogous study has been carried out by Kunau and coworkers with one of the S. *cerevisiae* peroxisome assembly mutants, *pasl,* previously isolated by this group. The *pasl* gene product is a hydrophilic 117-kDa protein with a putative ATP-binding site. The findings of these two groups represent major achievements.

Peroxisomal membranes

Studies on peroxisomal membranes have greatly benefitted from the development by Lazarow and coworkers of a one-step sodium carbonate extraction of purified organelles. This procedure effectively releases soluble and peripheral membrane proteins and yields a preparation consisting of membrane lipids and integral membrane proteins (IMPs). The total organelle proteins are generally recovered either in the soluble or in the membrane fraction. However, their enzymatic activity is frequently lost upon carbonate treatment. In such cases, their distribution can be established by an alternative technique involving Triton X-114 phase separation. Studies on the osmotic behavior of isolated peroxisomes and the lack of latency of urate-, D-aminoacid-, and L-α-hydroxyacid oxidase indicated that sucrose and the substrates for these three peroxisomal oxidases easily permeate through the membrane of the isolated organelles.

Direct permeability measurements with seemingly intact rat liver peroxisomes revealed them to be permeable to small solutes including glucose, sucrose, urea, acetate, methanol, uric acid, and the cofactors for fatty acid oxidation, i.e. NAD^+, CoA, ATP, and carnitine. Evidence was obtained from reconstitution experiments with peroxisomal IMPs in phospholipid vesicles that the high permeability of the peroxisomal membrane is caused by the presence of a pore-forming protein.

The authors suggested that a 22-kDa IMP that is unique to peroxisomal membranes is the most likely candidate for the pore-forming activity. It was also suggested that peroxisomal metabolism is not regulated at the level of membrane transport and that the concentrations of unbound substrates and cofactors are of the same order of magnitude in the peroxisomal matrix and the cytoplasm. This concept is hard to reconcile with the existence of a proton gradient across the peroxisomal membrane as observed in several yeasts by ^{31}P NMR measurements and immunocytochernistry. A proton-translocating ATPase associated with the cytosolic face of the peroxisomal membrane of yeasts was held responsible for the acidic internal pH of 5.8 to 6.0 of peroxisomes.

The presence of an ATPase in yeast peroxisomes has recently been challenged, however. The existence of a pH gradient across peroxisomal membranes in organisms other than yeasts has not yet been established. However, the association of a cytosol-oriented ATPase activity with rat liver peroxisomes and evidence for the involvement of an ATPase in the transport of at least one of the substrates for DHAP-AT across the peroxisomal membrane have been reported. These results are all i with the notion that the leakiness of isolated peroxisomes represents an isolation artifact.

Integral mammalian peroxisomal membrane proteins

Table 3 lists the IMPs of known molecular weights reported to be present i mammalian peroxisomes. The 76-kDa peroxisomal long-chain acyl-CoA synthetase in rat liver appears to be identical to the microsomal and mitochondrial enzyme. The 69/70-kDa protein has been detected in peroxisomal membranes from rat liver and human liver and fibroblasts. It was also found in liver and fibroblasts from Zellweger patients. The 69/70-kDa protein in rat liver peroxisomes extends into the cytoplasm, and evidence has been presented that 68-, 41/42-, and 28-kDa membrane proteins are derived from the 69/70-kDa polypeptide by the action of endogenous proteases during isolation of peroxisomes. Masters and coworkers have reported that mouse liver peroxisomes contain one IMP with an M_r of 68,000, and another with an M of 70,000.

The latter appeared to be identical to the bifunctional protein, which is generally not regarded as an IMP. It is known to be poorly soluble and to remain partly associated with the membrane fraction even after carbonate extraction. A cDNA for the 69/70-kDa peroxisomal IMP from rat liver has now been cloned and sequence. Its C-terminal region shows strong sequence homology to a superfamily of ATP-binding proteins, many of which are involved in ATP-driven transport of small molecules or proteins through membranes.

The ATP-binding domain of the peroxisomal membrane protein was localized to the part extending into the cytoplasm, in line with a proposed role in ATP-driven transport into the organelle. Other peroxisomal IMPs for which a functional role has been proposed include 22-kDa and 35-kDa polypeptides. The 22-kDa protein has been detected in peroxisomes from rat and human liver and human fibroblasts and is also found in liver and peroxisomal ghosts in fibroblasts from Zellweger patients. As discussed in the previous section, this 22-kDa protein is believed to be involved in the permeability of peroxisomal membranes to small molecules. The protein consists 44% of hydrophobic amino acids, and appears to be embedded deeply in the membrane, in that proteolysis of intact peroxisomes produced a slightly smaller polypeptide of 21 kDa. Quite recently, the cloning and characterization of a cDNA encoding a unique peroxisomal membrane protein of 35 kDa that restores the biogenesis of peroxisomes and complements the defect of peroxisomal functions in a peroxisome-deficient CHO mutant was reported.

This protein, denoted PAF-1, contains two putative hydrophobic membrane-spanning regions and is highly accessible to proteinase K, suggesting that a considerable part is exposed on the outer surface of peroxisomes. Peroxisomal IMPs of as-yet-unknown function include a 57-kDa protein in rat liver, a 53-kDa protein present in liver, and fibroblasts from both controls and Zellweger patients, and 36-kDa, 26-kDa and 15-kDa proteins in rat liver. The 76-, 69/70-, 26-, and to a lesser extent the 57-kDa protein increase in rodent liver upon administration of peroxisome proliferators, whereas the 36 and 22-kDa proteins are much less affected, thus causing a change in membrane polypeptide composition upon proliferation.

The 69/70-, 36-, 26-, and 22-kDa polypeptides were all shown to be synthesized as the mature proteins on free polysomes. Other proteins located to peroxisomal membranes include DHAP-AT, alkyl-DHAP synthase, and very long-chain acyl-CoA synthetase, but the polypeptides responsible for these activities have yet to be identified.

Integral plant glyoxysomal membrane proteins

In comparison to mammalian peroxisomes, less work has been reported on the IMP composition of glyoxysomes. In carbonate pellets of glyoxysomes isolated from Castor bean endosperm, 22 bands were detected by SDS-PAGE.

Remarkably, 12 of these were reported to be glycoproteins. However, these results have not been confirmed by others and have been the subject of some controversy.

A possible relationship with IMPs from mammalian peroxisomes has not been investigated. It is noteworthy that the abundant, presumably pore-forming, 22-kDa protein from mammalian peroxisomes does not seem to be present in glyoxysomes. In this context it may be relevant to note also that isolated glyoxysomes do not seem to be freely permeable to small molecules as evidenced by latencies for malate synthase, hydroxyacyl-CoA dehydrogenase, and malate dehydrogenase of 98, 85, and 71%, respectively.

Integral yeast peroxisomal membrane protein

Research on the IMPs of yeast peroxisomes is essentially confined to C. tropicalis and C. *boidinii.* Four major proteins of 44, 34, 29, and 24 kDa were found in C. tropicalis peroxisomal membranes, of which the 24-kDa entity was induced by growth on oleic acid. This pattern is significantly different from that observed in C. *boidind,* where the most prominent bands consisted of 120-, 100-, 47-, 32-, 3 1 -, and 20-kDa polypeptides. None of these appear to be glycosylated.

The 20-kDa protein and its mRNA could only be detected after a shift from glucose to methanol as C-source. Immunoelectronmicroscopy detected a predominant labeling over the peroxisomal membrane, but a significant fraction of the signal appeared in the matrix as well. This and the fact that a large proportion of the 20-kDa protein could be removed from the membrane by carbonate treatment suggested that it may be a peripheral membrane- or a matrix protein. In agreement with this notion, no obvious membrane-spanning region could be predicted from the base sequence of the isolated gene. In contrast to the 20-kDa protein, the 31/32- and 47-kDa proteins represented abundant components of the peroxisomal membrane in *C. boidinii,* regardless of whether oleic acid, D-alanine, or methanol was used to induce peroxisome proliferation. The 47-kDa protein could also be observed by immunoelectronmicroscopy in the small organelles that are present in cells grown on glucose.

The gene encoding this protein has been isolated, and its sequence predicts two membrane-spanning segments. It appears to be a novel protein with no sequence similarity to other proteins. The abundant presence of this protein regardless of the peroxisome inducer utilized suggests that it may be related to the structure, assembly, or some other general function of peroxisomes in yeast.

Chapter 8

Biochemistry of Inositol Phosphate

Berridge proposed a scheme in which phospholipase C (PLC) hydrolyzed phosphatidylinositol 4,5-bisphosphate [PtdIns $(4,5)P_2$] to yield inositol 1,4,5-trisphosphate [Ins$(1,4,5)P_3$] and diacylglycerol. However, the multiple sources of diacylglycerol, as well as the many intermediates, enzymes, and routes for formation of inositol phosphates, indicate that this signal transducing pathway is much more complex than described previously.

Many new reactions, enzymes, and inositol phosphate-binding proteins have been discovered, and molecular genetic approaches have begun to further define the structures and functions of the proteins involved. Rather than catalog all of the new information, I present selected topics that have yielded new insights for future directions of research. We do not discuss calcium ion mobilization in response to Ins $(1,4,5)P_3$ as this topic has been reviewed elsewhere. The main subjects to be discussed are

(a) mechanisms for triggering phosphatidylinositol turnover and inositol phosphate production;

(b) characterization of enzymes that metabolize inositol phosphates;

(c) pathways for metabolism of inositol pentaphosphates and inositol hexaphosphates;

(d) inositol phosphatebinding proteins, especially the Ins(1,4,5)P_3-binding protein; and

(e) the newly discovered 3-phosphate-containing inositol phospholipids.

There are three major myo-inositol containing lipids: phosphatidylinositol (Ptdlns), phosphatidylinositol phosphate [Ptdlns PI, and Ptdlns P_2. Ptdlns accounts for more than 80% of the total phosphatidylinositols. The pathway for biosynthesis of Ptdlns in the endoplasmic reticulurn was elaborated in the early 1960s. Ptdlns(4)P and Ptdlns(4,5)P_2 are synthesized sequentially by Ptdlns 4-kinase and PtdIns(4)P 5-kinase. Ptdins 4-kinase is membrane-associated in most tissues, and has been purified to homogeneity from bovine uterus, A431 cells, porcine liver, and rat brain.

The properties of the purified enzymes are very similar. Ptdlns(4)P 5-kinase has been found both in the soluble and particulate fractions of cell homogenates. A 53-kDa form of this enzyme has been purified from human erythrocyte membranes. There appear to be two distinct forms of Ptdlns(4)P 5-kinase in erythrocytes. Type II Ptdlns(4)P 5-kinase is distributed in both membrane and cytosol and has a molecular weight of 53,000.

A second form, Type I Ptdlns(4)P 5-kinase, is found only in the membrane fraction of erythrocytes and though not yet purified to homogeneity, appears distinct from the previously characterized 53-kDa Type 11 enzyme. Antibodies against the cytosolic Type 11 enzyme neither inhibit the activity of Type I enzyme nor detect Type I enzyme by immunoblotting. The reason for multiple isoforms and' the physiologic functions of each remain to be determined. Functional characterization of these kinases shows enzyme. Additionally the Type I Ptdlns(4)P 5-kinase phosphorylate that the Type I kinase has a lower K_m for Ptdlns(4)P than does the Type I Ptdlns(4)P in isolated erythrocyte membranes, whereas the Type II Ptdlns(4)P 5-kinase has no activity towards membrane Ptdlns(4)P.

Phospholipase C

PtdIns-specific PLC enzymes catalyze the hydrolysis of phosphatidylinositols to produce messenger molecules. There are multiple PLC enzymes in mammalian tissues as deduced from direct protein isolation and molecular cloning studies. Nine isoforms of PLC identified to date have been categorized into four groups, designated α, β, γ, and δ. None of the PLC isoforms contains a membrane-spanning sequence with the possible exception of PLCα.

The phospholipid substrates for these enzymes are all in membrane bilayers; therefore PLC must bind to membranes before hydrolyzing phosphatidylinositols. Reversible membrane binding of PLC is an attractive hypothesis to explain changes in Ptdlns turnover upon stimulation of cells by agonists, although such has not been demonstrated experimentally.

phospholipase C by GαQ

Numerous studies have demonstrated that GTP and its nonhydrolyzable analogues stimulate Ptdlns turnover in permeabilized cells, crude membrane fractions, or partially purified enzyme preparations. This has led to speculation that a guanine nucleotide-binding protein (G protein) is involved in coupling between receptors and PLC. Recently a new class of G protein, designated Gq, that activates the PLCβ1 isoform of PLC has been discovered. Pang & Sternweis isolated G_αq from rat brain by a novel strategy utilizing affinity chromatography on columns containing βγ subunits. Detergent extracts of brain membranes were allowed to bind to the affinity column in the presence of GDP. When the extract was passed very slowly over the column, the a subunits of the G proteins dissociated from endogenous βγ and bound to the matrix βγ subunits.

The α subunits of the G proteins were then eluted from the column by addition of the nonhydroiyzable GTP analog, $GTP_{\gamma}S$. $G_{\alpha}q$ remained bound to the column under these conditions because it exchanges GDP for $GTP_{\gamma}S$ very slowly, but eluted with aluminum fluoride. Aluminum fluoride binds to GDP on $G_{\alpha}q$ and thereby mimics GTP, causing dissociation of $G_{\alpha}q$ from the column. Through use of this method, two novel a subunits were obtained, $G_{\alpha}q$ and $G_{\alpha}11$. Limited sequence information indicated that these proteins correspond to two G protein α subunits that were cloned by Strathmann & Simon using a strategy based on the polymerase chain reaction (PCR).

Taylor *et al* have also isolated $G_{\alpha}q$ from bovine liver by a different strategy. These workers treated liver membranes with $GTP_{\gamma}S$ overnight prior to solubilization with detergents and in this way identified a fraction that could be separated from PLC and that stimulated PLC activity. The isolated G protein was shown to be Gq by immunoblotting with antipeptide antibodies specific for Gq. The $G_{\alpha}q$ subunit specifically activates PLCβI and has no stimulatory effect on other isoforms of PLC tested, including PLCγ and PLCδ.

Taylor et al demonstrated a 30-fold stimulation of PLCβI by $G_{\alpha}q$ from approximately 10 μmol/min/mg protein to almost 300 μmol/min/mg protein. The activation was independent of calcium ion concentration. In similar studies, Smrcka *et al* found that a preparation containing PLCβ1 was markedly stimulated by Gq. However, in this case Gq appeared to lower the affinity of the enzyme for calcium ions. At 0. 1 μM free calcium, marked stimulation occurred (~30 fold), whereas at 20 μM calcium only a threefold stimulation occurred. These studies identify at least one isoform of PLC that is activated by a traditional G protein mechanism. Whether other PLC isoforms are similarly activated by other types of G proteins remains to be elucidated.

PLCγ by tyrosine phosphorylation

Whereas the direct activation $PLC\beta_1$ by Gq-type G proteins has been convincingly demonstrated, the mode of activation of PLCγ by tyrosine protein kinase receptors is not clear. PLCγ has been shown to be phosphorylated on tyrosine residues in cells responding to platelet-derived growth factor (PDGF) and epidermal growth factor (EGF). In cells containing EGF receptors and PLC-β, -δ, and -γ, only PLCγ is phosphorylated in response to EGF stimulation of intact cells.

Although it is clear that PLCγ is a substrate for receptor tyrosine protein kinases, an effect of the phosphorylation on the activity of the enzyme has not been demonstrated. In fact, the magnitude of phosphorylation of PLCγ on tyrosine is variable. For example, in A431 cells, which have large numbers of EGF receptors, up to 50% of PLCγ may be phosphorylated; in PDGF-stimulated Balb 3T3 cells less than 5% of PLCγ complexes with receptors and most is not phosphorylated, and much of the tyrosine-phosphorylated PLCγ is found in the cytosol. Kim *et al* have demonstrated that EGF receptor kinase phosphorylates PLCγ stoichiometrically in vitro on tyrosines 472, 771, 783, and 1254.

However, phosphorylation had no effect on PLC activity as measured using $Ptdlns(4,5)P_2$ as substrate at a variety of calcium ion concentrations. In contrast, Nishibe *et al* immunoprecipitated PLCγ from homogenates of

A431 cells, phosphorylated it in vitro with EGF receptor kinase, and found approximately threefold increased PLC activity in the immunoprecipitates. They showed that treatment of these precipitates with a protein tyrosine phosphatase (T-cell phosphatase) both decreased tyrosine phosphate and reduced PLC activity by 50%.

However, Goldschmidt-Clermont *et al* found that PLC-/ phosphorylated by the EGF receptor had only 25% increased enzyme activity using Ptdlns(4,5)P_2 as substrate. Factors present in the immunoprecipitates may account for the differences in the observed results. Kim *et al* have transfected mutant forms of PLCγ with substitution of phenylalanine for tyrosine into NIH $3T_3$ cells and studied the effects on PLC activity measured in cell homogenates versus inositol phosphate production in response to stimulation of intact cells with PDGF.

Thus the former monitors the effects of tyrosine phosphorylation on enzyme activity, whereas the latter measures the effect of tyrosine phosphorylation on the coupling of phosphatidylinositol turnover to PDGF stimulation in intact cells. When either tyrosine 783 or 1254 was mutated to phenylalanine, cellular phosphatidylinositol turnover in response to PDGF was either totally ablated or attenuated. Homogenates from both of these mutants expressed the same PLC activity in vitro as those from cells transfected with native PLCγ. This study shows that phosphorylation of PLCγ is necessary for increased phosphatidylinositol turnover in cells stimulated with PDGF.

The authors speculate that tyrosine phosphorylation may reverse some inhibitory factor not apparent in enzyme assay of homogenates. One cytosolic protein that may affect PLCγ activity is the actin-binding protein profilin. Goldschmidt-Clermont showed that profilin binds Ptdlns (4,5)P_2, inhibiting the ability of the lipid to be hydrolyzed by unphosphorylated PLCγ. In the presence of 50 μM profilin, Ptdlns(4,5)P_2 hydrolysis by PLCγ was reduced fivefold. However, when PLCγ was phosphorylated by the EGF receptor, no inhibition was observed. Interestingly, profilin did not inhibit the action of PLCβ on Ptdlns(4,5)P_2.

One problem with this study is that the assay conditions used involved substrate vesicles containing inhibitory phospholipids, including phosphatidylserine and/or phosphatidylcholine. Thus, PLC activity in the absence of profilin was only ~0.1% of that obtained under optimal conditions. Another compound that may affect PLCγ activity is glucosphingolipid. Shayman *et al* have shown that reduction of cellular glucosphingolipids with an inhibitor of glucosylceramide synthatase, threo-I-phenyl- 2-decanoyl-amino-3-morpholino-1-propanol, increases phosphatidylinositol turnover in response to bradykinin or GTγS in MDCK cells. However, it is not known which PLC isozymes are present in MDCK cells.

Inositol polyphosphate 5-phosphatase

There are multiple isoforms of inositol polyphosphate 5-phosphatase enzyme distributed between soluble and particulate fractions of various tissues. The 5-phosphatase acts on three (1,3,4,5)P_4 and clns(1:2,4,5)P_3 (3). Two forms of cytosolic 5-phosphatase have been isolated from human platelets (47, 48). Type I enzyme has an apparent molecular weight of 45,000 and requires Mg^{2+}. The K_m of Type I 5-phosphatase for Ins(1,3,4,5)P_4 is 0.5 μM, much lower than that for Ins(1,4,5)P_3 (K_m = 7.5 μM). The V_{max} for Ins(1,3,4,5)P_4 is only 2% of that obtained using Ins(1,4,5)P_3 as substrate.

Consistent with these findings, low concentrations of Ins(1,3,4,5)P_4 (I μM) inhibit the breakdown of either Ins(1,4,5)P_3 or clns(1:2,4,5)P_3 in homogenates. This implies that cellular levels of $InsP_3$ metabolites may be sustained under conditions where Ins(1,3,4,5)P_4 is formed. Type I 5-phosphatase from human platelets is phosphorylated by protein kinase C, and this phosphorylation is associated with increased 5-phosphatase activity. Thus, protein kinase C action tends to reduce $InsP_3$ levels and terminate signalling. Type II 5-phosphatase from human platelets hydrolyzes the same substrates as the Type I enzyme, although the kinetic parameters are different (K_m values for Ins(1,4,5)P_3 = 24 μM, and for Ins(1,3,4,5)P_4 = 7.5 μM).

The Type II enzyme has a molecular weight of 75,000 and is not phosphorylated by protein kinase C. Recently, a cDNA encoding human platelet Type II 5-phosphatase has been isolated by screening a placental γgt11 cDNA library for β-galactosidase fusion proteins that bind to Ins(1,3,4,5)P_4. The sequences derived from the expression clone were used to screen human erythroleukernia and megakaryocyte cell cDNA libraries. The composite cDNA isolated consists of 2381 bp and predicts a protein that includes the amino-terminal amino acid sequence (19 residues) obtained from the platelet 75-kDa Type II 5-phosphatase.

The cDNA predicts that the mature enzyme contains 635 amino acids (M_r = 72,891). The recombinant protein expressed in Cos-7 cells has the same intrinsic 5-phosphatase activity as the platelet 5-phosphatase as determined by immunoblotting. All of the 5' cDNA sequences isolated to date maintain an open reading frame 5' to the predicted amino-terminal sequence of the platelet Type II enzyme. Thus it is possible that the platelet 5-phosphatase is formed by proteolytic processing of a larger precursor.

The full length of the transcript based on Northern blotting is 4.4 kilobases. Antibodies directed against recombinant Type II 5-phosphatase immunoprecipitate

Type II enzyme from homogenates of platelets and do not cross-react with Type I enzyme, further supporting the idea that two distinct isoenzymes are present. There is no sequence homology between Type II 5-phosphatase and other cloned proteins of the phosphatidylinositol pathway, including PLC, Ins(1,4.5)P_3 3-kinase, inositol monophosphate phosphatase, inositol cyclic 1:2 phosphate 2-phosphohydrolase, inositol polyphosphate I-phosphatase, or the Ins(1,4,5)P_3-binding protein. In addition no homologies were found when comparisons were made to sequences in the Genbank data base.

Sequence analysis of the predicted protein identified several potential phosphorylation consensus peptide sequences for cGNIP-dependent protein kinase, proline-dependent kinase, casein kinase 1, casein kinase 11, mammary gland casein kinase, and glycogen synthetase kinase-3. Consistent with the lack of phosphorylation of Type II 5-phosphatase by protein kinase C in vitro, there are no putative protein kinase C phosphorylation sites. 5-phosphatase enzymes have been partially characterized in several other tissues.

The characteristics of these enzymes suggest that additional isoforms of 5-phosphatase are very likely present. Thus molecular weights in excess of 100,000 have been noted as well as a near inability to hydrolyze Ins(1,3,4,5)P_4 by some isoforms. In particular, two distinct 5-phosphatase enzymes have been identified in rat brain and bovine brain. 5-phosphatase activities have also been characterized in extracts from macrophages, pancreas erythrocytes, and liver. In hepatocytes. It appears that the particulate 5-phosphatase may be assymetrically localized between canilicular and sinusoidal membranes.

INS (1, 4, 5)P_3 3-Kinase

Another enzyme that utilizes Ins(1,4,5)P_3 as a substrate is Ins(1,4,5)P_3 3-kinase, a widely distributed soluble enzyme that converts Ins(1,4,5)P_3 to Ins(1,3,4,5)P_4 in the presence of Mg^{2+} and ATP. The 3-kinase enzyme phosphorylates Ins(1,4,5)P_3 but not cIns(I:2,4,5)P_3.

The Ins(1,4,5)P_3 3-kinase has low K_m for Ins(1,4,5)P_3 (0.2 to 1.5 μM). Thus the 3-kinase may compete effectively with the inositol polyphosphate 5-phosphatases for Ins(1,4,5)P_3, since the K_m, for the latter enzymes are higher (7-25 μM). The native enzyme is composed of two catalytic subunits of 53 kDa plus calmodulin. Physiological concentrations of Ca^{2+} stimulate Ins(1,4,5)P_3 3-kinase activity via the calmodulin subunit. Ins(1,4,5)P_3 3-kinase may also be regulated by phosphorylation. Addition of phorbol esters to intact platelets or to a malignant T-cell line causes a twofold increase in 3-kinase activity. In hepatocytes, the enzyme is activated by stimulation of protein kinase *A* and protein kinase *C* together. In contrast, when isolated Ins(1,4,5)P_3 3-kinase is Phosphorylated in vitro with protein kinase C, it is inhibited. In vitro phosphorlation of the isolated enzyme by cAMP-dependent protein kinase increased the apparent enzyme activity.

The site of phosphorylation by protein kinase A was serine 109, whereas multiple sites of phosphorylation by protein kinase C were observed including serine 109 and serine 175. Results of other studies suggest additional mechanisms for regulation of Ins(1,4,5)P_3 3-kinase. Cellular Ins(1,4,5)P_3 3-kinase activity was increased 6- to 8-fold in cytosolic extracts prepared from vSRC-transformed cells. Striking developmental changes in the activity of 3-kinase occur in rat brain, where enzyme activity increases 15-fold from birth to 13 weeks of age. Ins(1,4,5)P_3 3-kinase has been isolated from rat and bovine brain.

The enzyme is susceptible to proteolysis by calpain, which apparently accounts for the finding of multiple molecular-weight forms ranging from 32 to 53 kDa. A cDNA encoding Ins(1,4,5)P_3 3-kinase has been isolated from a rat liver cDNA library and predicts a protein of 449 amino acids with a molecular weight of 49,853.

The protein has no sequence relationship to other enzymes of inositol phosphate metabolism. There are four short sequences from different regions of the molecule that are 20 to 50% identical to sequences in the Ins(1,4,5)P_3-binding protein. NIH 3T_3 cells and CCL39 hamster fibroblasts stably transfected with the 3-kinase cDNA overexpress InsP_3 3-kinase 16- to 18-fold. These cells display lower levels of Ins(1,4,5)P_3 and higher levels of Ins(1,3,4,5)P_4 after stimulation with thrombin, PDGF, or bombesin, consistent with increased kinase activity. Blunted calcium mobilization was observed, presumably due to lower levels of Ins(1,4,5)P_3.

The previously proposed role for Ins-(1,3,4,5)P_4 in calcium mobilization is not supported by this study, since in the 3-kinase-transfected cells the elevated Ins(1,3,4,5)P_4 levels had no effect in promoting calcium ion mobilization. Further study of the physiology of these cells will be required to uncover a function for Ins(1,3,4,5)P_4 or other "downstream" metabolites elevated in these cells.

Inositol polyphosphate 1-phosphatase

Inhorn *et al* discovered a phosphatase in calf brain homogenates that hydrolyzes Ins(1,3,4)P_3 and Ins(1,4)P_2 to Ins(3,4)P_2 and Ins(4)P, respectively. The enzyme was designated inositol polyphosphate I-phosphatase, although it does not hydrolyze other I-phosphate-containing inositol polyphosphates. Subsequently, others have found this enzyme in brain and liver.

The enzyme has been isolated from calf brain and is a monomeric protein of 44 kDa. The K_m for Ins(1,4)P_2 is 4 to 5 mM and that for Ins(1,3,4)P_3 is 20 μM. The enzyme requires magnesium ions and is inhibited at physiological calcium ion concentrations. Rabbit polyclonal antisera against I-phosphatase inhibits hydrolysis (greater than 95%) of Ins(I 4)P_2 in extracts of bovine tissues, suggesting that I -phosphatase accounts for the majority of Ins(1,4)P_2 hydrolysis.

Lithium ions inhibit both lns(1,3,4)P_3 and Ins(1,4)P_2 hydrolysis uncompetitively with a K_i of 0.3 μM and 6.0 μM, respectively. The differences in K_i for lithium are consistent with the uncompetitive mode of inhibition, i.e. the inhibitor reacts only with the enzyme-substrate complex. The inhibitory potency of lithium for Ins(1, 3,4)P_3 hydrolysis suggests that inhibition at this step may account for the pharmacologic action of lithium ions used to treat psychiatric disorders, since lithium levels in patients are approximately I μM. Bovine brain inositol polyphosphate I-phosphatase has been digested with cyanogen bromide, and trypsin and peptide fragments were isolated and sequenced.

The sequence was used to design degenerate oligonucleotide primers that were used to isolate a 1572-bp cDNA with an open reading frame of 400 amino acids, corresponding to a predicted molecular weight of 43,980. The predicted I-phosphatase amino acid sequence is not similar to any other sequences in Genbank with the exception of slight homology to inositol monophosphate phosphatase. In one region of the molecule, 9 of 20 amino acids are identical and in another segment 6 of 15 amino acids are identical. It is of interest that these similar regions are also conserved in three proteins found in *Escherichia coli.* These proteins have pleiotropic effects when mutated but have no other known functions. It is possible that they are signalling molecules of *E. coli.*

The cDNA for the I-phosphatase has been expressed

in *E. coli*. The recombinant protein has the same specific activity, affinity for substrates, and inhibition by lithium ions as native inositol polyphosphate I-phosphatase. These results suggest that the activity of the native protein does not require posttranslational modifications.

The cDNA for bovine brain inositol polyphosphate I-phosphatase was used to screen a human umbilical vein endothelial cell cDNA library to obtain a 1.75-kilobase cDNA encoding human inositol polyphosphate I-phosphatase. The predicted amino acid sequence of the human I -phosphatase is 82% identical to that of the bovine brain clone.

Inositol polyphosphate 4-phosphatase

Inositol polyphosphate 4-phosphatase converts Ins(1,3,4)P_3 to Ins(1,3)P_2 and Ins(3,4)P_2 to Ins(3)P. The enzyme does not require metal ions, is not inhibited by Li^+, and does not hydrolyze the 4-phosphate from other inositol polyphosphates. Barisal et al have purified this enzyme 3400-fold from half brain-soluble extract.

The isolated enzyme has an apparent molecular mass of 110 kDa as determined by gel filtration. On SDS polyacrylamide gel electrophoresis the protein migrates at 105 kDa, suggesting that it is monomeric. The inositol polyphosphate 4-phosphatase has apparent K_m values of 40 and 25 μM for Ins(1,3,4)P_3 and Ins(3,4)P_2, respectively. The maximal velocities for these two substrates are 15 to 20 μM of product/minute/mg protein. Ins(1,3,4)P_3 is a competitive inhibitor of Ins(3,4)P_2 hydrolysis with an apparent K_i of 27 μM, implying that the same active site is involved in hydrolysis of both substrates.

The ratio of inositol polyphosphate I-phosphatase to inositol polyphosphate 4-phosphatase activity in hydrolyzing their common substrate Ins(1,3,4)P_3 varies among tissues. Only 5 to 20% of Ins(1,3,4)P_3 is utilized by inositol polyphosphate 4-phosphatase in various tissues, except in brain, where approximately 70% is metabolized by 4-phosphatase. These observations suggest that Ins(1,3)P_2 or a further metabolite plays a role in neuronal function.

Inositol polyphosphate 3-phosphatase

Inositol polyphosphate 3-phosphatase catalyzes the hydrolysis of the 3 -position phosphate bond of inositol 1,3-bisphosphate [Ins(I,3)P_2] to form inositol I-phosphate and inorganic phosphate. Two isoforms of this enzyme designated Types I and 11 have been isolated from rat brain. The Type I 3-phosphatase consists of a protein doublet that migrates with a molecular mass of 65,000 upon SDS polyacrylamide gel electrophoresis.

The molecular weight of this isoform upon size exclusion chromatography is 110,000, suggesting that the native enzyme is a dimer. The Type II enzyme consists of equal amounts of a molecular weight 65,000 doublet and a molecular weight 78,000 band upon SDS polyacrylamide gel electrophoresis. This isoform has a molecular weight of 147,000 upon size exclusion chromatography, indicating that it is a heterodimer.

The 65,000-Da subunits of the two forms of 3-phosphatase appear to be the same based on comigration on SDS polyacrylamide gels and peptide maps generated with Staphylococcus aureus protease V8 and trypsin. The peptide map of the 78,000-Da subunit was different from that of the 65,000-Da subunits. The Type II 3-phosphatase catalyzes the hydrolysis of Ins(1,3)P_2 with a catalytic efficiency 1/19th of that measured for the Type I enzyme, suggesting that the presence of the 78-kDa subunit decreases activity. Both isoforms of 3-phosphatase enzyme also hydrolyze phosphatidylinositol 3-phosphate [Ptdlns(3)PI to form phosphatidylinositol and inorganic phosphate.

The Type II enzyme is more active than Type I enzyme in hydrolyzing Ptd Ins(3)P. Thus far this is the only example in inositol phosphate metabolism of an enzyme that hydrolyzes both a lipid and corresponding water-soluble substrate.

Inositol Monophosphatase

The other enzyme of inositol phosphate metabolism that is inhibited by Li^+ in vitro is inositol monophosphatase. This enzyme was initially designated as inositol I -phosphatase, but since it hydrolyzes phosphate groups of all inositol monophosphates with the exception of inositol 2-phosphate, its name has been changed to inositol monophosphate phosphatase. Inositol monophosphate phosphatase requires Mg^{2+} for activity and other divalent ions, i.e. Ca^{2+} and Mn^{2+}, are competitive inhibitors.

Hallcher & Sherman demonstrated that Li^+ inhibits inositol monophosphate phosphatase uncompetitively with an apparent K_i of 0.8 μM; thus this enzyme is another potential target for the action of lithium as a therapeutic agent. Inositol monophosphate phosphatase has been purified to homogeneity from bovine and rat brain. The enzyme (M_r = 55,000) is a dimer of identical subunits. A cDNA encoding inositol monophosphatase was recently isolated and predicts a protein of 277 amino acids.

The recombinant protein has been expressed in *E. coli* and found to have the same activity and inhibition by Li^+ as native monophosphatase. The protein is not homologous to any other proteins in the Genbank data base with the exception of inositol polyphosphate I-phosphatase and the bacterial proteins mentioned above. Modification of monophosphatase by the arginine-specific reagent phenylglyoxal results in enzyme inactivation, suggesting that an arginine is in the active site of this enzyme. One mole of arginine is modified per mole of protein in the process of enzyme inactivation.

Inositol (1 : 2 cyclic) phosphate 2-phosphohydrolase and inositol cyclic phosphates

The metabolism of inositol cyclic phosphates is distinct from that of the more widely studied noncyclic counterparts. They are produced by PLC and subsequently metabolized by distinct enzymes. The existence of inositol cyclic phosphates in vivo has been documented in a wide variety of cells and tissues. It is not clear what determines the ratio of cyclic versus noncyclic phosphates produced upon stimulation of cells with various agonists.

In general, prolonged stimulation of cells or tissues tends to produce higher levels of cyclic phosphates, in part because these metabolites are metabolized slowly. For example, inositol polyphosphate 5-phosphatase hydrolyzes clns(1:2,4,5)P_3 very poorly compared to Ins(1,4,5)P_3, which may explain the accumulation of cyclic trisphosphate in cells. A 4-phosphatase in bovine brain hydrolyzes clns(1:2,4)P_2 and requires magnesium ions. This phosphatase differs from inositol polyphosphate 4-phosphatase. Inositol cyclic (1:2) phosphate 2-phosphohydrolase (cyclic hydrolase) is the only enzyme known to hydrolyze the cyclic phosphate bond, and this enzyme only utilizes inositol cyclic (1:2) phosphate [clns(1:2)P] as a substrate.

Since all inositol cyclic phosphates are finally metabolized by this enzyme, it is likely that it regulates cellular levels of these molecules. Both Zn^{2+} and Ins(2)P are potent inhibitors, while Mn^{2+} and acidic phospholipids stimulate the enzyme. Cyclic hydrolase also catalyzes the hydrolysis of a second substrate glycerophosphoinositol to form glycerol and inositol 1-phosphate. As in the case of hydrolysis of inositol cyclic phosphates, the glycerophosphoryl derivatives with additional monoester phosphate groups in drolase. Cyclic hydrolase has been isolated from human placenta and characterized and shown to be identical to lipocortin III. Lipocortin III is one of eight related proteins that bind lipids and calcium ions but have no previously known physiological functions, and have recently been renamed *annexins*. In order to investigate possible functions for inositol cyclic phosphates, Ross et al have transfected a cDNA encoding cyclic hydrolase into $3T_3$ cells.

$3T_3$ cells that overexpress cyclic hydrolase (3-10-fold) have decreased cellular levels of clns(1:2)P, indicating that alterations in the cellular levels of the enzyme directly alter cellular clns(1:2)P levels. Cells with increased cyclic hydrolase activity grow to a lower density at confluence (4-fold) than untransfected cells or cells transfected with vector alone. Similar results were obtained when the growth properties of several different cell types with varying endogenous cyclic hydrolase activity were examined. Those cells with high levels of cyclic hydrolase had correspondingly low clns(1:2)P levels and grew to a lower density at confluence than cells expressing low levels of enzyme (4-fold range).

Cyclic hydrolase levels have also been found to be lower within a given cell line during active cell growth than when cells are confluent. These studies suggest that cyclic hydrolase is antiproliferative and could be an example of an antioncogene. Whether cyclic hydrolase can reverse or prevent the transformed phenotype remains to be determined. While cyclic hydrolase is important in determining cellular clns(1:2)P levels, another way to vary this metabolite is through alterations in PLC activity. Camilli et al have shown that an essential virulence factor in the intracellular pathogen *Listeria monocytogenes* is a PLC enzyme. This enzyme produces only clns(1:2)P and diacylglycerol from PtdIns. Thus it is possible that overproduction of either or both of these metabolites is required to produce disease.

Other enzymes of inositol phosphate metabolism

Several other enzymes involved in $InsP_4$ metabolism have been partially characterized. Ins(1,3,4,6)P_4 is formed by a soluble 6-kinase that metabolizes Ins(1,3,4)P_3. The 6-kinase is widely distributed and is not affected by either calcium or lithium ions. A 3-phosphatase that converts Ins(1,3,4,5)P_4 to Ins(1,4,5)P_3 has been described in cell extracts. It has been suggested that this enzyme may actually be involved in $InSP_5$ and $InsP_6$ metabolism, since these substances are potent inhibitors of this 3-phosphatase.

Inositol pentaphosphate and hexaphosphate metabolism

The metabolism of inositol phosphates containing 5 and 6 phosphate esters is not fully understood. Further, it appears that there are differences among organisms in the particular isomers of $InsP_5$ present. There are six possible $InsP_5$ isomers, most of which have been found in various plant species and Dictyostelium. In animal cells, the major $InsP_5$ isomer is Ins (I,3,4,5,6)P_5, although other isomers including traces of (D, L)Ins (1,2,4,5,6)P_5 and (D,L)Ins (1,2,3,4,5)P_5 have been found in HL60 cells and NG 115 40IL-C3 cells.

Thus, Ins(1,3,4,5)P_4 is isomerized to Ins(1,3,4,6)P_4 via sequential action of inositol polyphosphate 5-phosphatase and Ins(1,3,4)P_3 6-kinase. Ins(1,3,4,6)P_4 is converted to Ins(1,3,4,5,6)P_5 by a 5-kinase. Ins(1,314,5,6)P_5 is further metabolized to three compounds: $InsP_6$, Ins(3,4,5,6)P_4, and Ins(1,4,5,6)P_4. The mass of $InsP_5$ and $InsP_6$ in HL60 cells has been measured as 35 μM and 50 μM, respectively. These levels are approximately 100-fold greater than those of other inositol phosphate metabolites. In HL60 cells stimulated with formyl-methionyl-leucyl-phenylalanine (FMLP), the levels of $InsP_3$ and $InsP_6$ rise within two minutes to 50 μM and 60 μM, respectively.

Thus, these compounds appear to undergo accelerated synthesis in response to agonists. In studies of $InsP_3$ and $InsP_6$ turnover, as measured by changes in [^{3}H] inositol incorporation into these compounds in response to agonists, conflicting results have been obtained, suggesting increased, decreased, or no change in turnover. Much of the variation in these results may be accounted for by the fact that isotope incorporation can reflect either changes in mass, specific activity, or a combination of the two.

Many investigators assume that labelling of tissue-culture cells with [^{3}H]inositol for 48 to 72 hours yields steady-state labelling wherein all metabolites of inositol have the same specific activity. This appears not to be the case in some instances and is never proven without mass measurements. In HL60 cells labelled with [^{3}H]Inositol for 48 hours, the specific activity of Ins(1,3,4,5)P_4, Ins(1,3,4,6)P_4, and $InsP_5$ was similar, yet upon stimulation with FMLP there was a rapid rise in specific activity of these compounds, thus indicating that they were not in equilibrium with precursor inositol phosphate metabolites.

The specific activity of (D,L)Ins(3,4,5,6)P_4 was approximately half that of $InsP_5$ in unstimulated HL60 cells, implying that this metabolite turns over very slowly. Upon stimulation with FMLP, labelling of (D,L)Ins(3,4,5,6)P_4 increased rapidly, suggesting accelerated turnover of this metabolite. Similar results were found in bombesinstimulated AR4-2J cells. The pathway for metabolism of these compounds in avian erythrocytes appears to be quite different. Stephens & Downes have analyzed $InsP_4$ and $InsP_5$ metabolism by labelling erythrocytes with [^{3}H]Inositol and $^{32}PO_4$. They analyzed the incorporation of $^{32}PO_4$ into various monoester phosphates of Ins(1,3,4,5,6)P_5 and Ins(3,4,5,6)P_4 in a protocol similar to that used by Cunningham *et al* to determine pathways of 3-phosphate containing inositol phospholipid metabolism. In this analysis the most highly labelled phosphate is the last added to the molecule.

In fact, the overall scheme they propose is Ins(4,6)P_2 → Ins(3,4,6)P_3 → Ins(3,4,5,6)P_4→ Ins(I,3,4,5,6)P_5. This pathway predominates despite the fact that avian erythrocytes have the enzymes to form both Ins(1,3,4,5)P_4 and Ins(1,3,4,6)P_4. It will be interesting to analyze the InsP_3 and InsP_5 isomers of animal cells by this same strategy. The functions of InsP_4, InsP_5, and InsP_6 in signalling remain a mystery.

InsP_5 and InsP_6 have been proposed to be extracellular agonists that stimulate neuronal excitability and reduce heart rate and blood pressure. It is unclear, however, how these compounds could exit cells. InsP_5 does reduce the affinity of hemoglobin for oxygen in erythrocytes of birds. Proteins that bind InsP_4, InsP_5, and InsP_6 have been identified recently, as described below.

Inositol phosphate—binding proteins

The action of Ins(1,4,5)P_3 in calcium ion mobilization has been validated through identification and isolation of an Ins(1,4,5)P_3-binding protein that serves as a calcium channel across various membranes. The isolated Ins(1,4,5)P_3-binding protein has a subunit polypeptide chain of 260 kDa upon SDS polyacrylamide gel electrophoresis, but a molecular weight of approximately I million upon size exclusion chromatography, thus implying that the native molecule is a tetramer. Cross-linking studies of the isolated protein also suggest a tetramer. The binding protein had a K_d for Ins(1,4,5)P_3

of ~ 10 nM. Other inositol phosphates displace bound Ins(1,4,5)P_3 poorly (1000-fold higher concentration). The Ins(I 4,5)P_3 -binding protein has been reconstituted into lipid vesicles and shown to directly mediate the transfer of calcium across the lipid bilayer.

Photo-affinity labelling and direct binding studies have shown that the protein contains an ATP-binding site and further that ATP affects the flux of calcium. The distribution of the Ins(1,4,5)P_3-binding protein in brain is highly enriched in purkinje cells, where 10 to 100 times greater concentration of this protein is present than in other cells. Mikoshiba and coworkers noted that the enrichment in purkinje cells and other properties of the Ins(1,4,5)P_3-binding protein suggested that it was likely to be similar or identical to a protein that they had previously characterized as abundant in purkinje cells and nearly lacking in cerebella of mice with various mutations in which purkinje cells are deficient or absent. These workers utilized monoclonal antibodies to isolate a cDNA encoding the Ins(1,4,5)P_3-binding protein.

The full-length transcript is approximately 10 kilobases and encodes a protein of 2749 amino acids with a molecular weight of 313,000. The predicted protein contains an ATP-binding motif and several putative transmembrane domains near the carboxyl terminus of the molecule. The majority of the protein is thought to reside in the cytoplasm of the cell with the carboxyl terminus in the lumen of the endoplasmic reticulum (i.e. outside the cell). These authors concluded that there might be as many as seven membrane spanning domains contained within the molecule. Based on studies with monoclonal antibodies, they concluded that the number of membrane spanning domains was odd and therefore the amino terminus and carboxyl terminus were on opposite sides of the membrane. They and others also noted a striking homology between the Ins(I 4,5)P_3 -binding protein and the ryanodine receptor. The homology is particularly strong in the carboxylterminal membrane- spanning domains.

One puzzling difference is that the ryanodine receptor has an even number of membrane- spanning domains and the carboxyl terminus is on the cytoplasmic face of the membrane. Both the Ins(1,4,5)P_3-binding protein and ryanodine receptor appear to be tetramers of similar structure as determined by electron microscopy, and both are involved in the movement of calcium across cell membranes. A partial cDNA encoding the purkinje cell Ins(1,4,5)P_3-binding protein had been isolated previously by obtaining cDNA clones encoding proteins uniquely expressed in mouse purkinje cells. This was achieved by using library of normal mouse cerebellar cDNA subtracted with cDNA sequence from the cerebella of mice with mutations causing purkinje cell degeneratior.

It is interesting that the predicted mass of the monomeric Ins (1,4,5)P_3-binding protein is 313 kDa (plus some carbohydrates since the molecule is a glycoprotein), but by SDS polyacrylamide gel electrophoresis the molecular weight appears to be 260 kDa. Recently a cDNA for rat brain Ins(1,4,5)P_3-binding protein was reported. Several different forms of Ins(1,4,5)P_3-binding protein cDNA have been identified. Nakagawa et al found five different subtypes- in various tissues during mouse brain development that they postulate arise by alternatively spliced exons. They used an S1 nuclease protection assay of RNA from different tissues and from mouse brain at different times after birth. They found that a "short-form" RNA with a deletion of 15 amino acids predominates in adult brain and accounts for 20-80% of the mRNA in other adult tissues. Danoff *et al* found the same short form in rat brain by a PCR assay, but curiously found the opposite developmental correlation in that the short form was in fetal brain while the long form predominates in adult tissues.

The significance of the different forms is unknown, and the lack of similarity of distribution in closely related species may suggest that the variant forms are not functionally important. Alternatively, the PCR assay may not accurately reflect quantities of mRNA in various tissues. While most of the physiological studies have suggested that calcium release is primarily from endoplasmic reticulum, localization of the Ins(1,4,5)P_3-binding protein by immunohistochemical methods suggests a potentially wider distribution of calcium-mobilizing sites.

Ross *et al*, utilizing immunoelectron microscopy of rat cerebellar sections, have shown that the Ins(1,4,5)P_3 -binding protein is distributed over much of the endoplasmic reticulum, but is also present over most of the nuclear membrane and in some areas of the cis golgi. This localization has suggested a possible nuclear site for calcium mobilization, which might mediate cellular responses involving cell growth or proliferation. Recently Ins(1,4,5)P_3 was shown to mediate calcium release from isolated nuclei. Evidence for other inositol phosphate-binding proteins has been obtained. A distinct Ins(1,3,4,5)P_4- binding protein (K_d = 5 nM) has been solubilized and partially purified from porcine and rat cerebellar membranes.

Donie et al have purified a porcine cerebellar Ins(1,3,4,5)P_4-binding protein approximately 20,000-fold. The purified protein has a mass of 42,000 daltons as estimated by SDS polyacrylamide gel electrophoresis. It binds 0.2 mol of Ins(1,3,4,5)P_4 mol of protein. The Ins(1,3,4,5)P_4-binding protein shows an approximately 100 times greater affinity for Ins(1,3,4,5)P_4 than for Ins(1,4,5)P_3, the converse of that observed for the Ins(1,4,5)P_3-binding protein. Theibert *et al* recently isolated two Ins(1, 3,4,5)P_4- binding proteins from rat brain that show similar binding specificities to that found by Donie & Reiser; however, the molecular weights were quite different (182,000 and a dimer of 174,000 and 84,000).

The specificity of the various $InsP_4$-binding proteins for particular $InsP_4$ isomers has not been defined, and some preparations bind $InsP_5$ quite well. Specific binding proteins for $InsP_6$ have also been identified. $InsP_6$ binds to isolated cerebellar membranes with an apparent K_d of 60 nM. Displacement by $InsP_5$ required approximately I µM. Binding sites were found in rat brain membrare fractions from mitochondria, myelin, and synaptosomes and represented at least 1% of membrane protein. An apparently different $InsP_6$-binding protein has been purified from rat brain membranes by affinity chromatography on an Ins(1,3,4,5)P_4 derivative affinity column. It differs in that it binds $InsP_5$ almost as well as $InsP_6$. $InsP_6$-binding sites have also been found in brain membranes by autoradiographic staining of brain sections using [^{3}H]$InsP_6$.

It was determined that $InSP_6$-binding sites were particularly enriched in cerebellum, choroid plexis, and in the nucleus of the solitary tract. This distribution mimics that of the regions of brain mediating pharmacologic activities of $InsP_6$. Definition of the functions of these proteins awaits further study.

Phosphatidylionsitol 3-phosphate pathway

Phosphatidylinositols containing phosphate esters in the 3 position of inositol represent a recently discovered pathway of phosphatidylinositol metabolism. These include Ptdlns(3)P, phosphatidylinositol 3,4-bisphosphate

[Ptdlns(3,4)P_2], and phosphatidylinositol 3,4,5-trisphosphate [Ptdlns(3,4, 5)P_3. Ptdlns(3)P is present in cells under all conditions and serves as a precursor for the polyphosphorylated 3-phosphate-containing phosphatidylinositols.

The last two mentioned compounds are formed transiently in many cells in response to agonists and growth factors. Whitman *et al* first identified Ptdlns(3)P in transformed fibroblasts and proposed that the 3-phosphate-containing lipids serve a function in controlling cell proliferation. However, they are also found in normal cells and in nongrowing cells such as neutrophils, platelets, and brain, indicating that they serve a function other than, or in addition to, stimulating cell proliferation. The 3-phosphate-containing phosphatidylinositols are present in very small amounts in cells. In fact the mass of these lipids has not been measured in any system. Based on incorporation of trace radiolabelled inositol or phosphate, they represent only I to 2% of the levels of the previously known non-3phosphate- containing phosphatidylinositol polyphosphates.

The 3-phosphate-containing lipids are not hydrolyzed by PLC enzymes. Given this lack of reactivity and the small amounts of these compound it is unlikely that they serve as precursors of water-soluble messenger molecules. The water-soluble inositol phosphates that would be derived these lipids via PLC action, namely Ins(1,3)P_2, Ins(1,3,4)P_3, and Ins(1,3,4,5)P_4, are all present in cells and are formed by other reactions as described above.

The levels of these metabolites in stimulated cells, based on estimates fro radiolabelling, are 10 to 50 times greater than the estimated levels of 3-phosphate-containing phosphatidylinositols. Therefore, it is likely that the 3-phosphate-containing lipids either function as lipids per se or, they are converted to other as-yet-unidentified compounds. Many agonists that evoke classical PtdIns turnover also stimulate the production of 3-phosphate-containing phosphatidylinositols. In quiescen NIH 3T3 cells or unstimulated platelets, it is possible to label Ptdlns(3)P with [^{3}H] inositol or $^{32}PO_4$, indicating that there is turnover or synthesis of Ptdlns(3)P in a basal state. After stimulation of platelets with thrombin, neutrophils with FMLP, or NIH $3T_3$ cells with PDGF, there is a rapid increase in labelling of both Ptdlns(3,4)P_2 and Ptdlns(3,4,5)P_3, suggesting that the formation of the polyphosphorylated lipids is the functionally significant event.

A question that remains to be addressed is whether all agonists that stimulate PtdIns tuınover activate both 3-phosphate- and PLC-mediated inositol lipid turnover. The pathway for formation of Ptdlns(3,4)P_2 and Ptdlns(3,4,5)P_3 is uncertain. Carpenter *et al* have isolated a PtdIns 3-kinase that phosphory-lates Ptdlns(4)P and Ptdlns(4,5)P_2 in vitro. However, labelling of intact cells suggests a different pathway. Cunningham *et al* labelled human platelets with $^{32}PO_4$ for brief periods and stimulated them with thrombin.

The relative specific activities of the phosphate group at each position of the 3-phosphate-containing phosphatidylinositols were determined by degrading the deacylated and deglycerated compounds with specific inositol phosphate phosphatases. As labelling reaches a steady state, all phosphates are equally labelled. Prior to a steady state, the order of phosphate addition is determined by the relative specific activity of each phosphate. In this analysis, the last phosphate added reaches steady state first and therefore is identified by demonstrating that it has the highest specific activity.

These experiments clearly indicated that the major pathway in thrombin-stimulated platelets is PtdIns → Ptdlns(3)P → Ptdlns(3,4)P_2 → Ptdlns(3,4,5)P_3. This pathway for synthesis was also found in NIH $3T_3$ cells by determining the specific activity of each phosphate in Ptdlns(3,4,5)P_3 after brief PDGF stimulation of $^{32}PO_4$-labelled cells.

The phosphate in position 5 was most highly labelled, followed by 4, then 3, and finally 1. Further support for this pathway comes from the discovery of a Ptdlns(3)P 4-kinase in homogenates of platelets, erythrocytes, and human erythroleukemia cells that forms Ptdlns(3,4)P_2 from Ptdlns(3)P. Stephens *et al* studied the pathway for Ptdlns(3,4,5)P_3 synthesis in FMLP-stimulated neutrophils. They labelled cells with [^{3}H] inositol and $^{32}PO_4$ and analyzed the relative specific activity of $^{32}PO_4$ in each position of Ptdlns(3,4,5)P_3. They found that the 3-position was most highly labelled and thus suggested that Ptdlns(3,4,5)P_3 arises from Ptdlns(4,5)P_2. It is difficult to reconcile this study with those of Cunningham *et al.*

The main differences in procedure are that Stephens et at used crude extracts as sources of the phosphatases to degrade the compounds and used a different cell type. Discounting possible errors because of the crude enzymes, it is possible that 3-phosphate-containing lipids may be synthesized by different routes under varied conditions in different cells. PtdIns 3-kinase has recently been isolated from bovine thymus. Two discrete types of 3-kinase enzyme were isolated. Type I is a dionomeric protein of 110 kDa, while Type II is a heterodimeric form with subunits of 110 kDa and 85 kDa.

Peptide mapping analysis indicated that the 10-kDa proteins were the same, while the 85-kDa protein was different. This suggests that the enzyme catalytic unit is the I 10-kDa protein and that the 85-kDa subunit is a regulatory unit. The specific activity of Type I enzyme was fivefold greater than that of the Type 11 PtdIns 3-kinase, indicating that the monomer is an active type and the heterodimer is less active.

The Ptdlns(3)P 3-phosphatase enzymes described above are analogous, since they show varied activity in that the Type I enzyme is less active than the heterodimeric Type II enzyme. The Type II PtdIns 3-kinase has also been isolated from rat liver. Recently, a cDNA encoding the 85-kDa regulatory subunit has been obtained by three different laboratories. The 85-kDa regulatory subunit of PtdIns 3-kinase is phosphorylated on both tyrosine and serine. This protein forms complexes with receptors that stimulate Ptdlns(3)P synthesis in a manner similar to that described above for PLCγ. PtdIns 3-kinase forms complexes with wild-type but not a mutant form of middle *T* antigen.

The mutant, which encodes phenylalanine in place of tyrosine at the major site of phosphorylation by $pp60^{csrc}$, retains the ability to form complexes with src, and activates its tyrosine kinase. The incidence and spectrum of tumors induced by cells with this mutant middle *T* antigen is reduced, implying that tyrosine phosphorylation is needed for recruitment of PtdIns 3-kinase into a complex that is in turn important for oncogenesis.

The significance of these complexes in activating PtdIns 3-phosphate containing phospholipid turnover is as yet uncertain. An additional uncertainty is the messenger role that could be served by the 3-phosphate-containing inositol phospholipids. Ptdlns(,4,5)P_2 has been shown to promote actin polymerization in vitro by causing dissociation of gelsolinactin complexes and association of profilin-actin complexes. Profilin inhibits actin polymerization by binding G-actin, while gelsolin severs actin filaments. Ptdlns(4,5)P_2 binds profilin with high affinity (K_d < 0. 1 μM) with a stoichiometry of 5 moles Ptdlns(4,5)P_2/mole profilin. Ptdlns(4,5)P_2 also binds to gelsolin. When platelets are stimulated by thrombin, there is rapid polymerization of actin, causing them to extend pseudopods and spread on surfaces.

These events occur at a time when Ptdlns(4,5)P_2 levels fall transiently and then return to basal levels. This pattern of change is not consistent with the proposed role for Ptdlns(4,5)P_2 in binding profilin to promote actin polymerization. Dadabay *et al* (200) found no correlation between changes in Ptdlns(4,5,)P_2 and actin polymerization in EGF-stimulated A431 cells. In contrast, Ptdlns(3,4,5)P_3 levels do rise rapidly upon thrombin stimulation, at least as measured by incorporation of $^{32}PO_4$. Perhaps Ptdlns(3,4,5)P_3 binds profilin or gelsolin to serve this proposed function.

Eberle *et al* have shown that the time course of actin polymerization and subsequent depolymerization in neutrophils in response to FMLP stimulation parallels the rise and fall of Ptdlns(3,4,5)P_3 in these same cells. When ligand binding is reversed with an antagonist, F-actin rapidly depolymerizes and ^{32}P-labelled Ptdlns(3,4,5)P_3 disappears. If these patterns of labelling actually reflect changes in mass, then Ptdlns(3,4,5)P_3 may control transient actin polymerization.

Mass quantities of Ptdlns(3,4,5)P_3 are needed to determine whether this molecule binds profilin as well or better than Ptdlns(4,5)P_2. It is possible that proliferative responses to growth factors require changes in actin polymerization and associations between actin and cellular cytoskeletal elements.

Chapter 9

Biochemistry of Respiration

A distinction is usually made between two forms of, or stages in, respiration, *Aerobic respiration* involves the release of energy by the reaction oxygen with carbohydrates which are completely broken down to carbondioxide and water. Anaerobic respiration, on the other hand, releases energy from carbohydrates by reactions which do riot involve oxygen from the air. An average adult men, who at least, utilises about 250 ml of oxygen and eliminates 200 ml of carbon dioxide per minute. However, these quantities become ten times during strenuous exercise.

The supply of the required amount of oxygen to the tissues and the elimination of CO_2 have been brought about by blood. The arterial blood supplies required oxygen to the tissues and takes up carbon dioxide from them and gets converted to venous blood. In turn, this enters the lungs where it gives up carbon dioxide and takes up oxygen from alveolar air to get reconverted in to arterial blood. The aqueous tension in alveolar air is about 48mm. Hg.

The compositions of atmospheric and expired air are given as follows: About 1/4 of the oxygen of the inspired air has passed into the blood and has been replaced in the expired air by equal amount of CO_2.

Partial pressure of gases

The atmospheric air contains nitrogen, oxygen, carbon dioxide and water vapour. The pressure exerted by a mixture of gases is equal to the sum of their partial pressures.

$$AP—PH_2O=P_{N2+}\ PCO_{2+}\ PO_2$$

Oxygen by Blood

When the venous blood inters the lungs, it contains 12.6 ml oxygen at a partial pressure (conventionally referred to as pO_2) of 40 mm. of Hg It gets separated from the alveolar air by only a thin endothelial lining of the capillaries and alveoli (not more than 1-2v thick). The alveolar air is having oxygen at a pO_2 of 100ml. Ibis makes oxygen to rapidly diffuse through the thin membrane into the blood, first in physical solution into plasma, later to be taken up by the hemoglobin of the erythrocyte.

The total surface of the alveoli happens to be so large (50 to 100 sq.meters) that during the short period of about 0.7 seconds the blood takes to traverse the lung capillaries, it attains equilibrium with alveolar air and becomes the arterial blood. Arterial blood is having an average of 19.6 ml /100, ml. at a pO_2 of 90 mm of Hg. At this pressure only 0.3 ml of oxygen is in physical solution while the rest is present in loose combination with hemoglobin as oxyhemoglobin.

$$Hb + O_2 \longleftrightarrow HbO_2$$

Hemoglobin is 97% saturated with oxygen under these conditions. The arterial blood now passes to the tissues and as it enters the capillary bed of the tissues, it again gets separated only by a thin endothelial lining of the capillaries from the interstitial fluid, where the pO_2 is only 30mm. The O_2 in physical solution rapidly diffuses out first and followed by O_2 released by the dissociation of HbO_2.

$$Hb + O_2 \longleftrightarrow Hb + O_2$$

By the time the blood traverses the capillary bed and becomes venous blood, it has lost about 7.0 nil of oxygen and the pO_2 decreases to 40mm. The hemoglobin is only 70% oxygenated. These changes involved in oxygen transport are closely interconnected with changes which are involved in CO_2 transport and resulting pH changes in blood.

Oxygen Dissociation Curves

If a solution of myoglobin (which is having only one heme per molecule) is exposed to pure oxygen and the per cent saturation of the solution with oxygen is measured, the relationship between oxygen tension and per cent saturation of myoglobin yields a rectangular hyperbolic curve. At the O_2 tension of arterial blood (about 100 mm. Hg.) myoglobin is nearly 95% saturated. The resting capillary O_2 tension is about 40mm Hg. at the arterial end. In the active muscle, it may decrease to as low as 20 mm. Hg. Even at such low oxygen tensions myoglobin can be over 80% saturated with oxygen.

The tissue oxygen tension during activity will become as low as 5 mm.Hg. and thus enough gradient exists for the myoglobin to supply oxygen to the tissues. It parts with about 12% of oxygen carried by it under these conditions. In hemoglobin, four subunits are there in the molecule—alfa$_1$, alfa$_2$ beta$_1$ and beta$_2$ chains.

The four subunits would display a 'cooperative' effect in transporting oxygen. Each alfa unit is more or less rigidly attached to a beta unit, but the attachment to the other beta unit is less rigid (i.e., the attachment of alfa$_1$ to beta$_2$ and alfa$_2$ (beta$_1$). Even these less rigid attachments exhibit alterations between oxyhemoglobin and reduced hemoglobin, In the oxygenated state, these bonds become more relaxed (R state) and in the reduced state, these bonds are more taut (T state) because of breakdown of small salt bridges between the units in the R state, and their reformation in the T state.

The $\alpha_2 \beta_2$ dimer exhibits slight rotation (about 15°) in relation to the $\alpha_1 \beta_1$ dimer during the conversion of the R form to the T form. In the R form, Fe' atom would be exactly in the plane of t lie porphyrin ring. In the T form, it is 0.07 nm out of the plane. Also, a valine residue projects into the heme pocket of the beta chain in the T form and blocks the entry of oxygen. The valine residue gets retracted away from the pocket in the R form because of the slight rotation and loosening of the salt bridges. Due to all these factors, the R form is having affinity for oxygen which is several hundred times that of the T form.

Formation of the R form would be favoured when one or two oxygen atoms are able to occupy the heme pockets of the alfa units. Thus, the uptake of one or two oxygen atoms by the alfa units triggers the conversion of the T form to the R form and is able to increase the affinity of the hemoglobin molecule to oxygen several hundred fold. The reverse set of reactions take place if the molecule loses oxygen.

The salt bridges get reformed, the $\alpha_1\beta_1$ unit rotates back to the preoxygenated state, pockets of the beta chains and the affinity to valine projects back into the heme pockets of the beta chain and the affinity to oxygen gets enormously reduced. Oxygen, therefore, rapidly gets dissociated from hemoglobin, and the reduced Tform Of hemoglobin gets formed. Due to this cooperativity in taking up oxygen by the four subunits of hemoglobin, the 0 2 dissociation curves for hemoglobin take an S-shape.

Temperature

(a) A rise in temperature decreases hemoglobin saturation.

(b) At 25°C, hemoglobin is 88% saturated but at 37°C, it is only 56% saturated. Therefore, hemoglobin gives up oxygen more readily while passing from high to low oxygen more readily while passing from high to low oxygen tension (as from lungs to tissues) in warm-blooded animals than in cold-blooded animals.

Electrolytes

At low oxygen tensions, oxyhemoglobin gives up oxygen more readily in the presence of electrolytes.

Effect of CO_2

(a) The influence of CO_2 on the shape of the dissociation curve is actually the effect of carbonic acid formation with the lowering of the pH of the environment.

(b) The increase in acidity facilitates the dissociation of oxyhemoglobin.

(c) The ability of CO_2 to shift the slope of the oxyhemoglobin dissociation curve to the right is known as the *Bohr effect.* This effect is often described as causing a shift of the P-50 to the right. P-50 is the partial pressure (mm. Hg.) at which hemoglobin is 50% saturated. 2, 3-biphosphoglycerate, a compound formed during glycolysis in the red cells, also causes a significant shift of the P-50 to the right.

Carboxyhemoglobin

1. Hemoglobin undergoes combination with carbon monoxide more readily than with oxygen (210 times as fast) to form cherry-red carboxyhemoglobin.
2. This reduces the amount of hemoglobin to carry oxygen.

3. The carbon monoxide in the inspired air is 0.02%, headache and nausea occur.
4. If case, the carbon monoxide concentration is only that of oxygen in the air (about 0. 1 % carbon monoxide), unconsciousness occurs in 1 hour and death in 4 hours.

Role of Diphosphoglycerate: It acts as an allosteric regulator of hemoglobin activity, it occurs in the red cell in equimolar concentration to hemoglobin. The affinity of DPG. to hemoglobin is much moie (about two times as much) than oxyhemoglobin. It binds allosterically hemoglobin molecules and brings about changes similar to oxygen binding in tissue capillaries. The DPG levels gets increased erythrocytes of people living at high altitudes.

CO_2 itself as an allosteric regulator by undergoing, combination reversibly with the $-NH_2$ of the n-terminal amino acids to carbamine hemoglobin is having a higher affinity for CO_2 than oxyhemoglobin. As the O_2 get lost from the oxyhemoglobin, CO_2 is taken up by the hemoglobin to form carbamino-hemoglobin.

Fetal respiration. The fetus gets supply of oxygen from blood, the pO_2 of which is quite low, Still, the fetal hemoglobin (HbF) is able to draw oxygen from placental blood, because the O_2 of dissociation curve of HbF is appreciably shifted to the left compared to adult hemoglobin. This is because of the low affinity of HbF to DPG. At a pO_2 of 30mm while HbA is only 33% saturated with oxygen, HbF is somewhat 58% saturated.

Oxygen toxicity: Exposure to oxygen under high pressure gives rise to some symptoms and may prove fatal. Pure oxygen is an irritant to the lungs and gives rise to pulmonary edema. Oversaturation of hemoglobin with oxygen will prevent the transport of carbondioxide, especially the isohydric phase. The—SH enzymes are also oxidized to the inactive -S-S- forms.

Transport of Carbondioxide in blood

Carbondioxide in solution forms carbonic acid which will lower down the pH to the to acid side. But actually there occurs only lowering of pH by 0.01 to 0.03 in the conversion of arterial to venous blood, though large amounts 6.6 ml/100ml blood) of CO_2 added to the venous Hood. As arterial blood enters to capillary bed in the tissues, it is having 48.2 vol% of CO_2 at a pCO_2 of 40 mm. This is separated by endothelial lining of capillary from the interstital fluid which has a pCO_2 of around 50 mm. This gradient is sufficient to allow CO_2 to diffuse from the tissue fluid in to the capillary blood.

The CO_2 first taken up as dissolved CO_2 but it is soon dealt with by other chemical mechanisms. The CO_2 content rises to 54.8 vols/100 ml by the time it reaches the veins as venous blood. In the lungs the reverse set of conditions prevail where the blood in the capillaries gets exposed to alveolar air with a pCO_2 of 40mm. Carbondioxide diffuses out into the alveoli by a reversal of the chemical processes through the stage of physical solution to with the 6,6 ml gained for restoring the original CO_2 content and pCO_2 of arterial blood.

Mode of transport of CO_2

1. *Physical solution:* The solubility of CO_2 in plasma and erythrocyte fluid at pCO_2 of 46 (venous blood) is 3.2ml/100ml and at pCO_2 of 40 (arterial blood), it is 2.7m1/100ml. This is responsible for transport of about 0.5 ml of CO, which is 78% of the 6.6 ml of additional CO_2 transported by venous blood.

2. *Chemical transport:* (i) Transport by *plasma proteins :* The CO_2 reacts with water to form carbonic acid, which in turn, dissociates to form the hydrogen and bicarbonate

ions. The hydrogen ion is taken up by the protein buffers and the, bicarbonate by the bases present in plasma.

$$H_2O + CO_2 \longleftrightarrow H_2CO_3 \longleftrightarrow H^+ + HCO_3$$

$$H^+ + \text{Protein} \longleftrightarrow \text{H. protein}$$

$$HCO_3^- + B^+ \longleftrightarrow BHCO_3$$

Other buffer systems such as the phosphate may also take up the hydrogen ion. This is able to account for 4% of the CO_2 transport (4 % of 6.6 ml), i.e., about 0.26 ml.

Carbonic anhydrase: This enzyme is able to facilitate the formation of H_2O_3 and its form bicarbonate. It occurs in the erythrocytes and accounts for the more than 90% of the HCO_3 in the R.B.C.

(ii) *Transport by hemoglobin of the erythrocytes :* This occurs by two methods :

(a) *Formation of carbaminohemoglobin*: Some of the free amino group of, hemoglobin readily combine with CO_2 in a reversible reaction to form carbaminohemoglobin

$$Hb\ NH_2 + CO_2 \longleftrightarrow Hb\ NH\ COO^-$$

Reduced hemoglobin is having a greater affinity for CO_2 than oxyhemoglobin. As the uptake of CO_2 and the release of O_2 go together, the conditions become highly favourable of carbaminohemoglobin formation. This accounts for 20% of the CO_2 transported (20% of 6.6 ml) i.e., 1.3 ml.

(b) *Isohydric transport of CO_2 by hemoglobin*: Quantitatively this is the most important and accounts for about 70% of the CO_2 transported (70% of 6.6ml i.e., 4.6 ml). Further, the transport is managed without the slightest change in pH on this account (hence isohydric)

Haldane effect. Binding of O_2 to hemoglobin displaces CO_2 from it. This is called Haldane effect.

Both hemoglobin and oxyhemoglobin behave as, weak acids. Between the two, oxyhemoglobin is stronger acid (pK = 2.4×10^7) and dissociates to yield more hydrogen ions than reduced hemoglobin (pK = 6.6×10^9). If the two hemoglobin are represented to indicate their acid nature as $HHbO_2$ and HHb, then the conversion of $HHbO_2$ to HHb which takes place during the passage of blood through the capillary bed of tissues will need addition of hydrogen ion to the weaker acid HHb formed.

During this phase the reactions are taking place simultaneously due to intake of CO_2 from the tissues. The HHb will take up the H^+ ion and the HCO_3^- is taken up by the bases (K^+ mainly in the cell). This is called as the isohydric transport of CO_2.

$$H_2O + CO_2 \longleftrightarrow H_2CO_3 \longleftrightarrow H^+ + HCO_3^-$$

Chloride Shift

1. CO_2 combines with water to form carbonic acid (H_2CO_3) mainly inside the red cell by the enzyme carbonic anhydrase present in the red cells.
2. The carbonic acid is then buffered by the intracellular buffers (phosphate and hemoglobin) combining with potassium.
3. Bicarbonate ion also returns to the plasma and exchanges with chloride which shifts into the cell when the tension of CO_2 increases in the blood.
4. When the CO_2 tension gets reduced, chloride leaves the cell and enters the plasma.
5. Under normal conditions the red cell is impermeable to sodium or potassium. But it is permeable to

hydrogen, bicarbonate and chloride ions and intracellular sources of cation (potassium) are indirectly available to the plasma by chloride (anion) exchange. This allows the carriage of additional CO_2 (as sodium bicarbonate) by plasma.

6. The CO_2 entering the blood from the tissues enters the red cells where it forms carbonic acid by carbonic anhydrase. Some of the carbonic acid returns to the plasma. The remainder combines with hemoglobin buffer to form bicarbonate which then returns to the plasma in exchange of chloride. The chloride gets neutralized by potassium in the red cells.

7. All of these reactions are reversible. At the lung, when the blood becomes arterial, chloride shifts back into the plasma, thereby liberating intracellular potassium to buffer the oxyhemoglobin and in the plasma, neutralizing the sodium which gets liberated by the removal of CO_2 during respiration.

Significance of N_2 in respiration: There exists no chemical mode of transport of nitrogen in blood. It occurs only in physical solution. At the pN_2 of 570mm 0.9 ml of nitrogen is dissolved per 100 ml. This remains the same in arterial and venous blood and in tissues. When the pressure gets increased as in diving under the sea, more nitrogen gets dissolved.

If in such an individual, there occurs a sudden decompression, the dissolved nitrogen will get liberated as bubbles of nitrogen gas into the blood and tissues. In blood this may being about embolism. In tissues, particularly nervous tissues, it will bring about severe pain due to pressure on the nerves. The individual has to be again kept in a compressed atmosphere and decompression should be carried out in a gradual manner to allow time for the nitrogen to be swept out through respiration.

Control of respiration

Respiratory centre: This is situated in the medulla oblongata and sends out impulses to the respiratory apparatus. Increase in pCO_2 and H^+ ion concentration are able to stimulate the respiratory centre directly. The rate and depth of respiration get increased till the excess CO_2 is blown out, thereby allowing the PCO_2 and H^+ return to normal. Decrease in pCO_2 and H^+ ion concentration depresses the centre and acts in a reverse way to conserve CO_2 and raise the H^+ ion concentration.

Chemoreceptors of the carotid and aortic bodies: These get stimulated by a decrease in pO_2 and increase in pCO_2 and H^+ion concentration of the arterial Wood. The impulses are carried through nerves to the respiratory centre which is reflexly stimulated. The main stimulus for chemoreceptor has been pO_2 decrease. There are certain perfluoro compound like perfluorodecalin which can dissolve gases like O2 and CO_2 proportional to their partial pressures.

The dissociation curve is linear. They are also nontoxic, when used of short periods. Hence, in conditions which need partial or complete replacement of erythrocytes, emulsions of this compound are injected intravenously to temporarily serve the function of respiratory exchange of gases.

Buffer Systems of the Blood

1. It is known that venous blood carries more CO_2 than arterial blood Therefore, the pH of venous blood is more acid than that of arterial blood by 0.0 1 0.03 units, i.e., pH 7.40 and 7.43 respectively.
2. The blood buffer consist of the plasma proteins, hemoglobin, bicarbonates and inorganic phosphates.

3. When CO_2 enters the venous blood, the small decrease pH is able to shift the ratio of acid to salt in all the buffer pairs. When the ratio is more of the acid, cations, become available to form additional bicarbonate. In this respect, the plasma and bicarbonates would play a minor role.

4. The buffering action of the plasma proteins in important as they release sufficient cations for the carriage of about 10% of the total CO_2.

5. The phosphates in the red cells we known to carry 25% of the total CO_2

6. The most important is buffering role of hemoglobin and oxyhemoglobin which carries 60% of the CO_2 of the whole blood.

At the lungs the formation of oxyhemoglobin from reduced hemoglobin releases hydrogen ions which undergo reaction with bicarbonate form, carbonic acid. The low CO_2 tension in the lung shifts the equilibrium towards the production of CO_2 which is continually eliminated in the expired air.

$$H^+ + HC_3^- \longleftrightarrow H_2CO_3 \longleftrightarrow H_2O + CO_2$$

In the tissues, the oxygen tension gets reduced and hence oxyhemoglobin dissociates delivering O_2 to the cells and reduced hemoglobin is formed. CO_2 produced by metabolism enters the blood, where it gets hydrated to HCO_3 which ionizes to form H^+ and HCO_3^-. Reduced hemoglobin acting as an anion accepts the H+ ions forming acid-reduced hemoglobin (HHb). Very little change in pH takes place because the newly arrived H^+ ions are buffered by formation of very weak acid. When the blood returns to the lungs, these H^+ ions are released because of the formation of a stronger acid (oxyhemoglobin) and the newly released H^+ ion gets promptly neutralized by HCO_3^-. This reaction has been found to be necessary for the liberation of CO_2 in the lungs.

Acid-Base Balance

Respiratory acidosis—This condition arises if there occurs the accumulation of H_2CO_3 in the blood under certain circumstances.

Respiratory alkalosis—This condition arises if the rate of elimination of CO_2 is excessive following the reduction of H_2CO_3 in the blood.

Metabolic acidosis—This happens in the deficit of bicarbonate without any change in H_2CO_3

Metabolic alkalosis—This occurs in the excess of bicarbonate.

Causes of disturbances in acid-base balance

Metabolic acidosis

1. It happens in uncontrolled diabetes with ketosis.
2. It happens in some cases of vomiting when the fluids lost are not acid.
3. It also happens in renal disease, poisoning by an acid salt, excessive loss of intestinal fluids (as in diarrhea or colitis) and excessive losses of electrolytes.
4. Increased respiration is an important sign of uncompensated acidosis.

Respiratory acidosis

1 This takes place in any disease which impaires respiration such as pneumonia, congestive failure, asthma.

2. It also occurs in the depression of the respiratory center (as by morphine poisoning),.
3. A poorly function respirator also gives rise to this condition.

Metabolic alkalosis

Causes

(a) Due to the ingestion of large quantities of alkali in the treatment of peptic ulcer.

(b) In the high intestinal obstruction (as in pyloric stenosis).

(c) After prolonged vomiting or after the excessive removal of gastric secretions containing hydrochloric acid.

(d) In Cushing's disease and during corticotropin or cortisone administration.

Symptoms

(a) Tetany is caused because of the decrease in ionized serum calcium.

(b) The respirations are slow and shallow.

(c) The urine becomes alkaline.

(d) Elevated blood bicarbonate.

(e) Deficiency of potassium.

Respiratory alkalosis

1. This is caused by hyperventilation.
2. It may occur in patients in hepatic coma.

The role of the kidney in acid-base balance

1. The non-volatile acids including lactic and pyruvic acids, hydrochloric acid, phosphoric acid and sulphuric acid get produced by metabolic processes. About 50 150 mEq. of the inorganic acids are eliminated by the kidneys in 24 hours. These acids are partially buffered with cation, largely sodium. In the distal tubules of the kidney some of the cations get reabsorbed and the pH of the urine fails.

2. Kidney can buffer acids and conserve fixed base in the production of ammonia from amino acids. The ammonia is used to neutralize acids when formed in excess.

3. In kidney disease, tubular reabsorption of sodium in exchange for hydrogen is poor and there occurs excessive retention of phosphates and sulphates. As a result, acidosis takes place.

Hypoxia

This term means oxygen deficiency in any condition of insufficiency of tissue oxidation process. Four types of hypoxia are known

1. Hypoxic hypoxia.
2. Anemic hypoxia.
3. Stagnant hypoxia.
4. Histotoxic hypoxia.

1. *Hypoxic hypoxia:* It is characterized by normal oxygen capacity but adminished oxygen tension in the arterial blood thereby resulting of varying degree of hemoglobin unsaturation.

(a) High altitude—Hypoxia is caused by the diminished oxygen tension in the atmospheric air and consequently in the alveolar air and bloodstream. This gives rise to respiratory alkalosis.

(b) Rapid, shallow respiration—Shallow breathing is due to insufficient oxygenation of blood. This happens in case of diseases.

(c) Congenital heart disease—In the congenital cardiac defect a portion of the blood may flow directly from the right to the left side of the heart without passing through the lungs. Because of this, mixture of aerated and non-aerated blood in the systemic circulation brings about hypoxic hypoxia.

2. *Anemic hypoxia:* It is characterized by a diminution in the oxygen capacity of arterial blood because of the decrease in the amount of functioning hemoglobin.

3. *Stagnant hypoxia:* It is due to circulatory insufficiency, the rate of blood flow through the tissues being retarded, with resulting increase in the percentage volume of oxygen removed from the blood in its passage through the capillaries.

4. *Histotoxic hypoxia:* In this condition the oxygen supply becomes normal in every respect but the degree of oxygen utilization by the tissues gets diminished.

Chapter 10

Biochemistry of Muscles

Muscle is the largest single tissue in a the human body, comprising somewhat less than 25% of body mass at birth, more than 40% of body mass in the young adult, and somewhat less than 30% in the aged adult. Muscle is only a pulling machine, not a pushing machine. Therefore, given muscle most be antagonized by another group of muscles or another fore such as gravity or elastic recoil.

There are three types of muscle tissues in the body:

— Striated (Voluntary) or skeletal muscle

— Non-striated (Involuntary) or smooth muscle

— Cardiac muscle.

Most skeletal muscle as the name implies, is attached to bones, and its contraction is responsible for the movements of the skeleton. The contraction of skeletal muscle is controlled by the somatic nervous system and is under voluntary, control. The movements produced by skeletal muscle are primarily involved with interactions between the body and the external environment. Smooth muscle surrounds hollow organs and tubes such as the stomach, intestinal tract, urinary bladder, uterus, blood vessels, and air passages to the lungs. It is also found as single cells distributed throughout organs (e.g., spleen) and in small groups of cells attached to the hairs in the skin or the iris of the eye. Contraction of the smooth muscle surrounding hollow organs may propel the luminal contents through the hollow organ (e.g., the intestinal tract), or it may regulate the flow of the contents through tubes by changing the tube diameter and thus the frictional resistance to flow (e.g., in blood vessels).

Smooth muscle contraction is controlled by the autonomic nervous system, hormones, and factors intrinsic to the muscle itself; it is not usually under direct conscious control. The third type of muscle, cardiac muscle, is the muscle of the heart, and its contraction propels blood through the circulatory system. Like smooth muscle, it is regulated by the autonomic nervous system, hormones, and factors intrinsic to the muscle itself; it is not usually under direct conscious control. The third type of muscle, cardiac muscle,. is the muscle of the heart, and its contraction propels blood through the circulatory system.

Like smooth muscle, it is regulated by the autonomic nervous system, hormones, and intrinsic factors in the cardiac muscle. Although there are significant differences in the structures, mechanical properties, and control mechanisms of the three types of muscle, the force generation mechanism is similar in all of them.

Skeletal Muscle Fibres

Skeletal muscle is composed of fibrils surrounded by electrically excitable membrane, the *sarcolemma*. The individual muscle fibre consists of a bundle of many myofibrils arranged in parallel. These are embedded by intramuscular fluid termed the *sarcoplasm*. The fluid contains glycogen, the high energy compounds ATP and phosphocreatine. and the enzymes of glycolysis. The functional unit of muscle is *sarcomere*. It exists along the axis of a fibril at distances of 2.5 μm. Alternating dark and light bands (A bands and 1 bands) are present in the myofibrils.

The central region of the A band (the H zone) is less dense than the rest of the band. The I band is bisected by a very dense and narrow Z line. The cross section of myofibril under electron microscope shows that myofibrils is constructed of two types of longitudinal filaments. One type confined to the A band (the thick filament) contains mainly the protein *mysoin*. These filaments are about 16 rim in diameter.

The other filament (the thin filament) lies 'In the 1 band but extends also into the A band. The thin filaments are smaller than those of myosin (about 6 nm in diameter). The thin filaments contain the proteins actin, tropomyosin and troponin. Each thin filament lies between 3 thick filaments. When muscle contracts, there is no change in the lengths of the thick c filaments or of the thin filaments, but the H zone and the 1 bands shorten. Thus, the arrays of interdigitating filaments must slide past one another during muscle contraction. The cross bridges generate and sustain the tension.

The tension developed during muscle contraction is proportionate to the, filament overlap and thereby the number of cross-bridges. Each cross-bridge head is connected to the thick filament via a flexible fibrous segment that can bend outward from the thick filament to accommodate the interfilament spacing.

Protein Muscles

The muscle fibrils are composed of

— Proteins 20%.

— Water 75%.

(3) Inorganic materials, certain organic extractives and carbohydrates, (glycogen and its derivatives) 5%.

Muscle proteins are characterized by their elasticity,

Myosin

— Myosin is the richly abundant muscle protein and it is a globulin.

— It is soluble in dilute salt solutions and insoluble in water,

— Its molecular weight is 5,00,000 containing 2 major chains and 4 light

— The enzyme *trypsin* cleaves myosin into two components—*meromyosins*—of unequal size. Therefore, they are termed as light and heavy meromyosins.

— It has the adenosine triphosphatase (ATPase) activity.

— Myosin binds act informing *actomyosin*. Light meromyosin does not combine with actin. But heavy meromyosin combines with actin.

— Heavy meromyosin is a rod-shaped protein attached to the two globular components of myosin. The rod portion can be splitt off of the globular region by the action of papain and the resulting portions are termed as HMM (heavy meromyosin) S-2 (the rod) or HMM S-1 (the globular portion). Each HMM SA possesses an active site for ATPase activity and a binding site for actin.

— The 4 light chains of myosin molecule are bound to the HMM S-1 fragments. The light chains function as modulators of ATPase activity.

— Actomyosin is formed by 3 myosin molecules with 1 actin molecule.

Actin

— Actin is a globulin of molecular weight 60,000.

— It is the major constituent of thin filaments in striated muscle.

— During the preparation of actin by extraction with solution of low ionic strength it is obtained as a molecular weight of 42,000 in a globular configuration called G-actin.

— G actin polymerizes to the fibrous form F-actin in the presence of Mg++ and ATP is hydrolysed to ADP and Pi is released. Hence, ATP must be added to set depolymerization of F-actin to G-actin.

— Actomyosin is formed by actin and myosin. But ATP dissociates to actin and myosin.

$$\underline{\text{Actin + Myosin} \qquad\qquad \text{Actomyosin}}$$
$$\text{ATP}$$

Molecular Mechanisms of Contraction

The lengths of neither the thick filaments nor the thin filaments change during shortening; rather, the two sets of filaments slide past each other. The following observations led to the Sliding-filament mechanism of muscle contraction. When a skeletal muscle fiber shortens, the width of the A band in a sarcomere remains constant. Since this width corresponds to the length of the thick filaments, these filaments do not change length during muscle shortening. In contrast, the widths of both the 1 band and the H zone decrease during shortening as the thin filaments move past the thick filaments.

The H zone decreases as the ends of the thin filaments from opposing ends of a sarcomere approach each other, and it disappears when these thin filaments meet in the center of the A band. The 1 band also decreases in width for the same reason, i.e., as more and more of the thin filament length overlaps the thick filaments.

During shortening the length from the edge of the H zone to the nearest Z line (the length of a thin filament) remains constant. Thus, the lengths of the thin filaments also remain constant during shortening. What produces the movement of these filaments? The answer is *the cross bridges*. When a cross bridge is activated, it moves in an arc parallel to the long axis of the thick filament, much like an oar on a boat. If, at this time, the cross bridge is attached to a thin filament, this swivelling motion slides the thin filament toward the center of the A band, thereby producing shortening of the sarcomere.

One stroke of a cross bridge produces only a small displacement of a thin filament relative to thick filament, angular movement of the bridge while attacked, detachment from the thin filament, reatattachment at anew location, and repetition of the cycle. Each cross bridge undergoes its own independent cycle of movement, so that at any one instant during contraction only about 50 per cent of the bridges are attached to the thin filaments, while the others are a., intermediate stages of' a cycle. Let us look more closely at the filaments and their cross bridges. Molecules of actin are arranged in two chains, helically intertwined to form the thin filaments. Myosin is a much larger molecule with a globular end attached to a long tail.

Approximately 200 myosin molecules comprise a single thick filament; the tails of the molecules lie along the axis of the filament and the globular heads extend out to the sides, forming the cross bridges. Each globular head contains a binding site able to bind to a complementary site on an actin molecule. The myosin molecules in the two halves of each thick filament are oriented in opposite directions, such that all their tail ends are directed toward the center of the fiber. Because of this arrangement, the power strokes of the cross bridges in the two ends of each thick filament are directed toward the center, thereby moving the thin filaments at both ends of the sarcomere toward the center of the sarcomere during shortening. In addition to the binding site for actin, the globular end of myosin contains a separate enzymatic site that catalyzes the breakdown of ATP (adenosine triphosphate) to ADP (adenosine diphosphate) and inorganic phosphate, releasing the chemical energy stored in ATP.

The splitting of ATP occurs on the myosin molecule before it attaches to actin, but the ADP and inorganic phosphate generated remain bound to myosin. The chemical energy released at the time of ATP splitting is transferred to myosin (M), producing an energized form of myosin (M*).

$$M + ATP \xrightarrow[\text{ATP breakdown}]{} M^*.ADP.P,$$

The subsequent binding of this energized myosin to actin (A) via a cross bridge triggers the discharge of the energy stored in myosin. With the resulting production of the force that causes movement of the cross bridge (the ADP and P_1 are released from myosin at this time):

$$A + M^*.\ ADP.P_1 \xrightarrow[\text{Actin binding}]{} AM^*.ADP.P_1 \xrightarrow[\text{Energy release and bridge movement}]{} AM + ADP.P_1$$

This sequence of energy storage and release by myosin is analogous to the operation of a mousetrap. Energy is stored in the trap by cocking the spring (by ATP splitting) and released by springing the trap (by binding to actin). This is, of course, only an analogy; the actual structural changes in myosin which energy storage and release are unknown. During contraction, the myosin cross bridge binds very firmly to actin, and this linkage, must be broken at the end of each bridge cycle; the binding of a new of ATP to myosin is responsible for breaking this link:

$$\text{A.M.+ ATP} \xrightarrow[\textit{AM dissociation}]{} \text{A + M. ATP}$$

Thus, upon binding (but not splitting) a molecule of ATP, myosin dissociates from actin. The free myosin bridge then splits its bound ATP, thereby reforming the energized state of myosin, which can now reattach to a new site on the actin filament and so on. Thus, ATP performs two distinct roles in the cross-bridge cycle: (1) The energy released from the splitting of ATP provides the energy for cross-bridge movement; and (2) the binding (not splitting) of ATP to myosin breaks the link between actin and myosin at the end of a cross-bridge cycle allowing the cycle to be repeated.

Many cross-bridge cycles occur during even the briefest contraction of a muscle fiber. The importance of ATP in dissociating actin and myosin at the end of a bridge cycle is illustrated by the phenomenon of rigor mortis (death rigor) the stiffening of skeletal muscles after death. Rigor mortis begins 3 to 4 h after death and is complete after about 12 h, but then slowly disappears over the next 48 to 60 h. Rigor mortis occurs because of the lack of ATP. In the absence of ATP the myosin cross bridges are still able to bind to actin, but the subsequent breakage of the link between them and the energizing of myosin requires ATP, Which is not produced in a dead muscle cell.

The thick and thin filaments become linked to each other by the immobilized cross bridges, producing the rigid connection of a dead muscle. In contrast, in a living muscle at rest, the myosin bridges are not bound to actin and the filaments readily slide past, each other when the muscle is passively stretched.

Muscle Energy Metabolism

The ATP required as the contrast energy source for the contraction relaxation cycle of muscle can be generated by glycolysis, oxidative phosphorylation, creatine phosphate, or two ADP molecules. The ATP stores in skeletal lived during contraction, providing energy probably for less than I second of contraction. *Phosphagens* such as creatine phosphate prevent the rapid depletion of ATP by providing a readily available high-energy phosphate, which is all that is necessary to reform ATP from ADP. Creatine phosphate is formed from and creatine at times when the muscle is relaxed and ATP demands are not so great.

The enzyme catalysing the phosphorylation of creatine is creatine phosphokinase (CPK), a muscle-specific enzyme with clinical utility in the detection of acute or chronic disorders of muscle. Skeletal muscle saroplasm contains significant glycogen sores. The release of glucose from glycogen is dependent upon a specific muscle phosphorylase enzyme. This enzyme is missing in a specific disorder of muscle (McArdle's disease), a form of glycogen storage disease. ATP is also available from oxidative phosphorylation in tissue, a process dependent upon a constant oxygen supply.

Muscles that have high oxygen demands as a result of sustained contraction (such as to maintain posture) have the ability to store oxygen in *myoglobin*. Because of the heme moiety to which oxygen is bound in *myoglobin* muscles containing myoglobin are red as compared to white skeletal muscle. Table 1 compares some of the properties of fast or white skeletal muscle with slow or red skeletal muscle.

Table 1. Characteristics of fast and slow skeletal muscle

	Fast Skeletal Muscle	*Slow Skeletal Muscle*
Myosin ATPase	High	Low
Energy utilization	High	Low
Colour	White	Red
Myoglobin contraction	No	Yes
Rate	Fast	Slow
Duration	short	Prolonged

Myoadenylate kinase, an enzyme present in muscle, catalyses the formation of one ATP molecule and one AMP from two ADP molecules. This reaction is shown in Fig. 9 coupled with hydrolysis of ATP by myosin ATPase during muscle contraction.

The relationships between these various sources of ATP and its consumption during muscle contraction are also depicted. In humans, skeletal muscle protein is the major nonfat source of stored energy. This explains the very large losses of muscle mass, particularly in adults, resulting from prolonged caloric undernutrition. The study of tissue protein breakdown in vivo is difficult because amino acids released during intracellular breakdown of proteins can be extensively reutilized for protein synthesis within the cells, or the amino acids may be transported to other organs where they enter anabolic pathways.

However, actin and myosin methylated following synthesis of their peptide bonds, producing *3-methylhistidine* (3-MeHis). During intracellular breakdown of actin and myosin, 3-MeHis is released and excreted into the urine. When labelled was administered to rats and humans, it was found that the urinary output of the methylated amino acid provides a reliable index of the rate of myofibrillar protein breakdown in the musculature of rats or human subjects. The fractional rate of muscle protein breakdown is Pot significantly different in the elderly a compared with young adults, but since muscle mass is less in the elderly, this tissue contributes less to the whole body protein breakdown that occurs aging in humans. As noted above, skeletal muscle is the major reserve of protein in the body. In addition, this tissue is highly active in the degradation of certain amino acids as well as in the synthesis of others.

In mammals, muscle appears to be primary site of catabolism of the branched chain amino acids. It oxidizes leucine to CO_2 and converts the carbon skeletons of aspartate, asparagine, glutamate, isoleucine, and valine into intermediates of the tricarboxylic acid cycle. The capacity of muscles to degrade branched chain amino acids increases 3, to 5 fold during fasting and in diabetes.

Biochemistry of muscles

Muscle also synthesizes and releases large amounts of alanine and glutamine. These compounds are synthesized utilizing amino groups that are generated in the breakdown of branched chain amino acids, and nitrogen is then transferred to x-ketoglutarate and to pyruvate by transamination. Glycolysis from exogenous glucose is the source of almost all of the pyruvate for synthesis of alanine.

These reactions constitute the so-called "glucose-alanine cycle" wherein alanine from muscle is utilized in hepatic gluconeogenesis while at the same time bringing amino groups to the liver for removal as urea. The carbon skeletons of the amino acids that are degraded in muscle and enter the tricarboxylic acid cycle in muscle are converted mostly to glutamine and to pyruvate, which itself is further oxidized or converted to lactate. It thus appears that in fasting or the postabsorptive state, muscle releases most amino a acids coming from net protein breakdown except for isoleucine, valine, glutamate, aspartate, and asparagine, which are used to contribute to the formation of glutamine, which itself is released for use by other tissues. For many years, it has been observed that working muscle releases ammonia. It is now known that the immediate source of ammonia in skeletal muscle is AMP, which is deaminated to IMP, catalyzed by adenylate deaminase. IMP may be converted back to AMP by reactions utilizing aspartate and catalyzed by adenylosuccinate synthetase and adenylosuccinase.

Muscle Fatigue

When a skeletal muscle fiber is continuously stimulated at a frequency that produces a maximal tetanic contraction, the tension developed by the fiber eventually declines, This failure of a muscle fiber to maintain tension as a result of previous contractile activity is known as muscle fatigue. Figure 9 illustrates several characteristics of the fatigue process. Fatigue is not limited to tetanic contractions but can occur with twitches if they are repeated frequently enough over an extended period of time.

On the other hand, low-frequency twitches can be repeated indefinitely in certain types of skeletal muscles with no signs of fatigue, Thus, the onset of fatigue and its rate of development depend on the type of skeletal muscle and the intensity of contractile activity. If after the onset of fatigue, a muscle is allowed to rest fora period of time, it can recover its ability to contract upon restimulation. The extent of recovery after a period of rest varies depending upon the type of previous activity.

A muscle will fatigue fairly rapidly if stimulated at a high frequency, but it will also recover rapidly from this fatigue. This is type of fatigue that accompanies short duration, high-intensity types of exercise, such as weight lifting. On the other hand, fatigue that develops more slowly with long-duration, low intensity endurance exercise, such as long distance running, requires much longer periods of rest often upto 24 h before complete recovery is achieved. It might seem logical that depletion of available ATP would account for fatigue, but it does not.

The ATP concentration it, fatigued muscle is only slightly lower than in a resting muscle. Indeed it is neither the supply of energy not the ability of the contractile proteins to generate tension that is responsible for rather, a failure of some step in excitation contraction coupling appears to be the most likely cause of fatigue, but which step and what brings about its failure is currently unknown. Failure of the appropriate levels of the cerebral cortex to send excitatory signals to the motor neutrons of a particular muscle call lead to its failure to contract. This type of *psychological fatigue* causes an individual to stop exercising even though the muscles themselves are not fatigued.

Collagen

Collagen, the major macromolecule of connective tissues, is the most common protein in the animal world. It provides an extracellular framework for all metazoan animals and exists in virtually every animal tissue. There are at least 5 distinct types of collagen in mammalian tissues thus, they exist as a family of sharing many properties. The most definitive property molecules is their triple helix, a coiled coil of 3 polypeptide subunits. Each polypeptide subunit or alpha is twisted into a left handed helix of 3 residue per turn.

Three of these left-handed helices are then wound to a right handed superhelix to form, a stiff rodlike molecule 1:4 nm in diameter and about 300 nm long.

These triple helical molecules unique to collagen are then associated bilaterally and longitudinally into fibrils. The arrangement of collagen fibrils involves longitudinal staggering slightly less than one-quarter the length of the tripple helix. Between the end of one triple helix and the beginning of the next is a that may provide a site for deposition of hydroxyapatite crystals in bone formation.

Collagen fibrils range from 10 to 100 rim in diameter and are visible by microscope as banded structures in the extracellular matrix tissues. The other striking characteristic of the collagen molecule is that every third residue in the triple helical portion of each alpha chain. Glycine is the only amino acid small enough the limited space available down the central core of the triple helical molecule;, the triple helical molecule consists of glycine residues provided by each of the 3 alpha subunits. This repeating structure can be represented by where X and V are amino acids other than glycine.

The hydroxyproline residues contribute additional stability to the collagen triple helix by forming more intramolecular hydrogen bonds mediated through extra water molecules. Collagen also contains 3-hydroxyprolone in some X positions and *5 hydroxyline in Y positions. The collagen triple helix is stabilised by multiple interchain cross-links* between lysyl and hydroxylysyl residues. The chemical nature of those cross-link is described below. Mature collage is a glycoprotein containing saccharides attached in O-glycosidic linkage to the hydroxylysine residues.

The Synthesis of Collagen

Collagen is air extracellular protein but is synthesized as an intracellular molecule that undergoes posttranslational modification before becoming a mature collagen fibril. Like all secreted proteins, the precursor of collagen is processed as it passes through the endoplasmic reticulum and Golgi complex prior to appearing extracellularly. The earliest collagen precursor is a *preprollagen* that contains a leader or signal sequence of approximately 100 acids as its amino terminus. Preprocollagen is generated by ribosomes attached to the endoplasmic reticulum. As the signal sequence penetrates into the vesicular space of the endoplasmic reticulum, the leader sequence is cleaved off and lie amino-terminal end of *procollagen* continues to protrude into the endoplasmic reticulum space. At this site *prolyl 4 hydroxylase* and *lysyl hydroxylase* act on proline or lysine residues, respectively, in the Y position of the (Gly-X-Y)n peptide.

A prolyl 3-hydroxylase acts on prolyl residues in the X position immediately preceding 4-hydroxyproline. in the Y position. The procollagen molecule contains at its amino terminus a 20,000 MW peptide and at its carboxy terminus a 30 to 35 thousand-MW peptide, neither of which is present in mature collagen. Both of these propeptides contain cysteine, residues. While the amino-terminal propeptide collagen forms only intrachain disulfide bonds, the carboxy terminal peptides form both intrachain and interchain disulfide bonds.

Following the formation of these disulfide bonds, the procollagen molecules assemble as the triple helix. After formation of the triple helix, further hydroxylation of prolyl and lysyl residues cannot occur. The glycosyltransferase activities that attach glucose or galactose to hydroxylysine residues also require that the procollagen alpha chains he nonhelical. Following this intracellular processing, the glycosylated procollagen molecule reaches the outside of the cell by way of the Golgi complex.

Extracellular procollagen aminoprotease and procollagen carboxy protease remove the amino-terminal and carboxy-terminal propeptides, respectively. The newly formed collagen molecules have approximately .1000 amino acids and spontaneously assemble into collagen fibrils that are indistinguishable from the mature fibrils found in tissues. These fibrils, however, do not have the tensile strength of mature collagen fibrils until they are *crosslinked by a series of covalent bonds.*

The extra cellular copper-containing enzyme lysyl oxidase oxidatively deaminates the E-amino groups of certain lysyl and hydroxylysyl residues of collagen, yielding reactive aldehydes. The aldehydes can form Schiff bases with amino groups of other lysines or hydroxylysines or even glycosylated hydroxylysines. These Schiff bases are chemically rearranged and provide stable covalent cross-links such as new peptide bonds or secondary amino bridges.

The aldehyde component derived from a hydroxylysine forms a more stable cross-link than does the aldehyde derived from a lysyl residue. Aldol bridges also provide intramolecular cross-links. The intracellular and extracellular processing of the collagen precursor molecule is summarized in Table 2.

Isotonic and Isomeric Contractions

During an isotonic contraction the distance the muscle 0 shortens, as a function of time, is recorded. The distance moved can he directly measured by a pen that is attached to a muscle (or to muscle lever) and leaves a trace on a moving strip of paper as the muscle contracts and relaxes. During an isometric contraction, the force (tension) generated by the muscle as a function of time is measured To measure an isometric contraction the muscle is attached at one end to a rigid support and at the other to a force transducer, which controls the movement of a recording pen in proportion to the force exerted. The terminology and methods of recording contractions described above apply to both single fibers and whole muscles.

Table 2. Intracellular (endoplasmic reticulum)

(1) Cleavage of signal peptide.

(2) 4-Hydroxylation of Y-prolyl residues.

(3) 3-Hydroxylation of X-prolyl where, Y = 4-hydroxyprolyl residue. Extracellular

(4) 5-Hydroxylation of Y-lysyl residues.

(5) Glycosylation of hydroxylysyl residues.

(6) Formation of intrachain and interchain SS bonds.

(7) Formation of triple helix procollagen.

Extracelluar

(1) Cleavage of NH_2-terminal propeptide (MW 20,000).

(2) Cleavage of COOH-terminal propeptide (MW 3035,000).

(3) Formation of immature collagen fibrils.

(4) Oxidation of lysyl, hydroxylysyl, glycosylated hydroxylysyl residue to aldehydes.

(5) Cross-linking of chains and helical molecules of fibrils via Schiff bases and aldol condensations

Twitch Contractions

The mechanical response of a muscle fiber to a single action potential is known as a twitch. In an isometric twitch there is an interval of a few milliseconds, the *latent period,* before the tension begins to increase. It is during this latent period that the process associated with excitation-contraction coupling are occurring. The time interval from the beginning of tension development to the peak tension is the *contraction time.* Not all skeletal muscle fibers have the same contraction times. Some fast fibres have contraction times as short as 10 ms, whereas slower fibers may take 100 ms or longer. The time from peak tension until the tension has decreased to zero is known as the *relaxation time.* Comparing an isometric twitch with an isotonic twitch in the same muscle fiber. At heavier loads the latent period is longer but the velocity of shortening, the duration of the twitch, and the distance shortened all decrease.

During the latent period of an isotonic contraction, the cross bridges begin to develop force, but actual shortening does not begin until the muscle tension slightly exceeds the load; therefore, the heavier the load, the longer the latent period. If the load on a fiber is increased, eventually a load will he reached that the muscle is unable to lift, the velocity of shortening will be zero, and the contraction becomes isometric.

Chapter 11

Biochemical Techniques

The technique is used to isolate and purify cellular organelles so that their structure and function can be studied in detail. Some of the techniques employed in biochemical research have been collected in this chapter. This will help the students to acquaint themselves with the terms and methods of practising biochemist.

Centrifugation separation techniques are based upon the behaviour of particles in an applied centrifugal field. The particles are normally suspended in a specific liquid medium, held in tubes or bottles, which are located in a rotor. The rotor is positioned centrally on the drive, shaft of the centrifuge, particles which differ in, density, shape or size can be, separated since they sediment at different rates in he centrifugal field, each particle at a rate which is proportional to the applied centrifugal field. Centrifugation techniques are two main types.

Preparative centrifugation techniques are concerned with the actual separation, isolation and purification of for example, whole cells, subcellular organelles, plasma membranes, polysomes, ribosomes, chromatin, nucleic acids, lipoproteins and viruses, for subsequent biochemical investigations very large amounts of material may be involved when harvesting microbial cells from tissue culture or plasma from blood. Relatively large amounts of cellular particles may also be isolated in order to study their morphology, con, position and biological activity. It is also possible to isolate biological macromolecules, such as nucleic acids and proteins, from preparations which have received some preliminary purification by for example techniques are devoted mainly to the study of pure or virtually pure macromolecules or particles. It is primarily concerned with the study of the sedimentation characteristics of biological macromolecules and molecular structures, rather than with the actual collection of particular fractions.

Such studies yield information from which the purity, molecular weight and shape of the material may he deduced, Since preparative centrifugation techniques are more commonly used in undergraduate studies, this chapter will concentrate on these techniques and deal only briefly with analytical centrifugation techniques.

Analytical ultracentrifuges are capable of operating at speeds approaching 70, 000 rev min^{-1} (500,000g) and consist of a motor; a rotor contained in a protective armoured chamber which is refrigerated and evacuated, and an optical system which enables, the sedimenting material to be observed so as to determine concentration distributions within the sample at any during centrifugation. Ultracentrifuges invariably work under refrigeration and a partial vacuum to prevent air friction cooking the cell components.

They can used to separate the *molecular* components of cells by a technique called *density gradient centrifugation.* This involves suspending disrupted cells in a solution of specific gravity, such as sucrose or *caesium chloride solution,* and carefully-chosen centrifugation times and speeds. In advanced analytical ultracentrifugcs, the photographic plate detector system is replaced all accurate electronic scanning system which can plot the concentration of the sample at all points ill the analytical cell at any particular different wavelengths of light can be selected enabling the separate movement of single components in a mixture of substances to be monitored provided they absorb light at different wavelengths.

Microscopy

Biochemical analysis is frequently accompanied by light and electron microscopic examination of tissue, cell or organelle preparations to evaluate the integrity of samples and to correlate structure with function. Microscopy serves two independent functions of enlargement *(magnification)* and improved resolution (the objects as separate entities). Light microscopes employ optical lenses to sequentially focus the image of objects whereas electron microscopes use electromagnetic lenses.

Light and electron microscopes may work either in a *transmission or scanning,,* mode depending on whether the light or electron beam either pass through the specimen and are diffracted or whether they are deflected by the specimen surface. Polarised light microscopes detect optically active substances in cells, e.g., particles of silica or asbestos in lull, tissue, or starch granules in amylplasts.

Light microscopes in *phase contrast* mode are often used to improve image contrast of unstained material, e.g., to test cell or organelle preparations for lysis. Changes in phase of emergent light are caused mainly by either diffraction or by changes in the refractive index of material within the specimen, or even by differences in thickness of the specimen. At their point of focus the converging light rays show interference resulting in either increases or decreases in the amplitude of the resultant wave *(constructive or destructive interference* respectively), which the eye detects as differences in brightness. Microscopes using visible light will magnify approximately 1500 times and have a resolution limit of out 0.2 p.m whereas a transmission electron microscope is capable of magnifying approximately 200,000 times and has a resolution limit for specimens of about 1 nm.

The excellent resolving power of transmission electron microscopy (TEM) is largely a function very short wavelength of electron accelerated under the influence of an applied electric field. (An accelerating voltage of 100 kV produces a 4×10^{-3} mm). Scanning electron microscopes (SEM) use a fine beam of electron to scan back and forth across the metal coated specimen surface.

The secondary electrons that are generated from this surface are collected by a scintillation crystal which converts each electron impact into a flash of light. Each light flash inside the crystal is amplified by a photomultiplier and used to build up an image on a fluorescent screen. The principal application of SEM is the study of surfaces such as those of cells. The resolution about 6 rim.

Optical Microscope

In the optical microscope a condenser produces a parallel beam of light to illuminate the specimen, and the images is focused and enlarged by objective and eyepiece lenses . The magnification of each lens is normally stamped on the side so that Combined with a x 40 eyepiece a total magnification of x 400, However, the amount of detail which can be, seen depends not only upon the magnification but also upon the microscope's ability to distinguish two points close together. If objects are too close, beyond the resolving power, they will blur together no Matter magnification is used. The resolving power of (resolution) of a microscope is given by Abbe's formula:

$$R = \frac{0.5\lambda}{n.\sin\ \alpha}\ \mu m\ (\text{or nm})$$

Where λ = the wavelength of radiation in (or nm)

n = the refractive index (RI) of the medium between the specimen and the first lens.

α = ½ angle of the aperture.

A small value for R clearly means a better resolution which, according to Abbe's formula, can be obtained by:

(i) Making λ smaller, e.g., using short wave (bluish) daylight instead of longer wavelength (Yellowish) artificial light.

(ii) Increasing n, by using *immersion oil* (RI = 1.5) to fill the gap between the objective and the specimen, instead of air (RI=1.0). This requires the use of special oil immersion lenses.

(iii) Making (x as close to 90° as possible, at which point the lens and specimen are in contact.

Most parts of a cell are more or less transparent. At high magnification only a small section of the object is in focus at one time, and a clear image can only be obtained if the specimen is thin. Consequently, the image often appears unnaturally flat. A three dimensional drawing can be constructed, however, by using serial sections. These are thin sections of a structure cut sequentially and viewed in turn.

The Electron Microscope

Electrons have two basic properties which together made possible the development of *electron microscopes (EMs)* during the 1930s.

(i) They are *charged particles* so they respond to *magnetic fields.* In EMs, powerful magnets are used to focus electrons in a similar way to the glass lenses which focus light rays in the optical microscope.

(ii) They have wavelike properties, this means they can be used to form an image on a fluorescent screen (the electrons are themselves invisible).

The wavelength of an electron is variable but always minute . It is given by the formula:

$$R = \frac{1.23}{E} \text{nm}$$

where E is the voltage through which an electron is accelerated. Clearly, the higher the voltage the shorter the wavelength and since Abbe's formula applies equally to both optical and electron microscopes, the best resolution is achieved at high voltages, However, the benefits obtained by using very short wavelengths are partly offset by the aperture in EMs 2α being much less then 1°, for technical reasons.

The nett result is that the resolution of the most powerful EMs is about 0.5 nm (0.0005 μm) or 400X greater than the best optical microscopes. This make a maximum practical magnification of about 500,000X. Electrons are easily scattered by air so EMs are usually operated at high vacuum. This means that electronmicroscopists are normally restricted to viewing dead, dried specimens.

A further difficulty is that at the high voltage

needed for the best resolution, electrons tend to pass straight through the specimen leading to loss of contrast. Contrast can be improved, however, by using heavy metal stains which impede or deflect the electrons and result in the formation of shadows. There are in fact two main types of EM: *transmission electron microscopes* (TEMs) and scanning electron microscopes.

In TEMs electrons are transmitted either through or past the specimens onto a fluorescent viewing, screen. In SEMs, electrons bounce off the heavy-metal C shadowed surface of a specimen, and the resulting image is viewed on a television screen SEMs-give by far the best images of surface features with a depth of field over 300x better than optical microscopes. Against this, resolution (10nm) and effective magnification (x20 000) are rather poorer than TEMs. Preparing material for electronmicroscopy is a skilled, time-consuming operation.

Despite these problems, or perhaps because of them, electronmicroscopists become ecstatic when the shadow of a platinum-plated replica from a dried-up slice of dead cell appears on their viewing screens. Whether what they are looking at bears any resemblance to the structure found in live cells, or is merely an aberration caused by specimen preparation *(an artefact)* has led to heated arguments.

Comparison of Optical and Electron Microscopes

A summary of the main features of both 0Ms and EMs is given in Table 1. As the table suggests, the two forms of microscopy are complementary. Whilst electron-microscopy reveals fine detail, the use of the optical microscope for rapid viewing of live specimens in colour remains unrivalled.

Ion probe analysis

Electron microscopes may be equiped to perform X-ray

spectrochemical analysis. When a specimen is irradiated by an electron beam, an electron may be displaced from an inner to outer orbital. When this happens the vacated orbital may be infilled by an electron from a higher energy orbital with the emission of an X-ray photon characteristic of the difference in energy levels on the two orbitals.

Table 1. Comparison of optical and electron microscopy

	Optical microscopy	*Electron microscopy*
1.	Much lower capital outlay and negligible running costs	High capital outlary and heavy running costs
2.	Most techniques can be easily acquired	High level technical skill required
3.	Sample preparation normally takes a few minutes/hours	Sample preparation often several days
4.	Live or dead specimens be used. Dehydration not always necessary	Dead, dried specimens are used
5.	Colour or black and white	Black and white only
6.	Poor at visualising surface	Good surface features (SEMs) or internal detail (TEMs)
7.	Poor Depth of focus	SEM depth of focus 300x better than optical microscopy
8.	Resolution 170 nm at best Magnification x 2000 at best	Resolution 0.5 nm at best. Magnification x 500 000 at best

The binding energy of the electrons in an orbital is related to the charge on the nucleus; hence each element produces its own characteristic emission spectrum. Emitted photons are usually detected by *energy-dispersive analysis,* using lithium drifted solid-state detectors. Each photon emitted reacts with silicon atoms to produce an electrical pulse proportional to the energy of the photon. An electronic pulse height analyser and microcomputer are used to process the spectral data which is normally displayed on a video monitor. X-ray spectrochemical analysis permits the measurement on ion distributions *in situ* in SEM and TEM by combining the high spatial resolution of electron microscopy with the ability to determine subcellular elemental composition.

Areas as small as 100 nm^2 can be analysed under optimal conditions and detection of any element with an atomic number greater than 10 is possible. Measurement is made in terms of total concentration of the element rather than free ion activity and in this sense the method is comparable to flame photometry though with a higher sensitivity.

Chromatography

Chromatography is relatively a new technique which was first invented by Michael Tswett, a Russian botanist in 1906 in Warsaw, for the separation of coloured substances into individual components. Since then, the technique has undergone tremendous modifications so that nowadays various types of chromatography are in use to separate almost any given mixture, whether coloured or colourless, into its constituents and to test the purity of these constituents.

The name chromatography (Greek *Chroma* colour and Graphy writing) means colour writing. Chromatography may be defined as the uniform percolation of a fluid through a column of more or less uniformly divided substance which selectively retards by whatever means certain components of the fluid (A.F.P Martin). In other words, chromatography may be defined as a technique for the, separation of the components of mixtures by a continuous, distribution of the components between two phases, one of which is moving past the other. 'The systems associated with this definition are as follows:

(i) A solid stationary phase with a liquid or gaseous mobile phase and

(ii) A liquid stationary phase with a liquid or gaseous mobile phase.

The system (i) gives rise to adsorption chromatography while the system (ii) to partition chromatography.

Theory of Chromatography

Chromatography is essentially based on the distribution of solutes between two solvents-a mobile and a stationary. The ratio of the amount of solute retained in one solvent (usually the stationary) to the amount of same solute in the other solvent (mobile) measured under the constant conditions of temperature is termed as the distribution ratio (D). This ratio has been found to depend upon the nature of the solute to be distributed between two solvents. The solvent phase in chromatography may belong theoretically to any one of three states of matter (solid, liquid and gases) but the most common employed is the liquid. As a result of this long chain amino acids migrate more rapidly than glycols and monosaccharides quicker than disaccharides. Thus, a substance may be characterised by the speed at which it gets migrated.

Classification of Chromatography: The moving phase may be ą liquid or a gas. Based on the nature of the fixed and moving phase, different types are as follows:

(a) *Adsorption Chromatography:* It is based difference in the absorption coefficients. The adsorbed components are then eluted by passing suitable (solvents) through the column.

(b) *Partition Chromatography:* It operates by mechanism analogous to counter-current distribution. The fixed phase may be a liquid strongly adsorbed on a solid which acts as support. In this case, the solute gets distributed between the moving liquid (solvent). The technique is called partition chromatography.

Paper chromatography is a special case of partition chromatography in which the adsorbent column is a paper strip.

(c) *Gas Chromatography:* When the moving phase is a mixture of gases, Chromatography or vapour phase chromatography (VPC).

A few chromatographic techniques are described below.

Paper Chromatography

Introduction: The credit for the present full-fledged status of paper chromatography in the realm of separation techniques goes to the Cambridge of workers, *A.J.P. Martin* and his coworkers *R. Consden, A.H. Gordon and Syngle.*

Principle: This technique is a type of partition chromatography in which the substances are distributed between two liquids, i.e., one is the stationary liquid (usually water) which is held in the fibres of the paper and called *stationary phase;* the others is the moving liquid or developing solvent and called the moving phase.

The components of the mixture to be separated migrate at different rates and appear as spots at different point on the paper. Originally paper chromatography was used to separate mixture of organic substances such as dyes and amino acids only. But now this method has been perfected to separate cations and anions of inorganic substances as well.

Procedure: In this technique, a drop of the test

solution is applied as a small spot on a filter paper and the spot is dried. The paper is kept in a close chamber and the edge of the filter paper is dipped into a solvent called *developing solvent*. As soon as the filter paper gets the liquid through its capillary axis and when it reaches the spot, of the test solution (a mixture of two or more substances), the various substances are moved by solvent system at various speeds. When the solvent has moved these substances to a suitable height (15-18 cm), the paper is dried and the various spots are visualised by suitable reagents called *visualising* reagents. The movement of substances relative to the solvent is expressed in terms of R F values, *i.e.,* migration parameters.

Migration parameter: The position of migrated spots on the chromatograms are indicated by different terms such as R_F, R_x, R_M and R_C These parameters are also qualitative and quantitative parameters, characteristic of a substance.

R_F : The R_F is related to the migration of the solute front relative to solvent front as:

$$R_F = \frac{\text{Distance travelled by the solute from the origin line}}{\text{Distance travelled by the solvent from the origin line}}$$

R_F is function of the partition coefficient. It is a constant for a given substance, provided the conditions of chromatographic system are kept constant with respect to temperature, type of paper, duration and direction of development, nature and the shape and the size of the wick used *(i.e.,* radial chromatography), the amount of liquid in the reservoir humidity, etc.

The R_F defines the movement of the substance relative to the solvent front in a given chromatographic system.

The R_F value of a substance depends upon number of factors which are:

(i) The solvent employed,

(ii) The medium used for separation, i.e., the quality of paper in case of paper chromatography,

(iii) The nature of the mixture,

(iv) The temperature, and

(v) The size of the vessel in which the operation is carried out.

Keeping the above factors constant, it is possible to compare the R_F values of different substances.

R_X: In some cases, the solvent front runs off the end of filter paper, the movement of a substance in such cases is expressed as R_X but not R_F.

$$R_X = \frac{\text{Distance travelled by the substance from the origin line}}{\text{Distance travelled by the standard substance } x \text{ from the origin line}}$$

Types of Paper Chromatography

1. *Descending Chromatography:* When the development of the paper is done by allowing the solvent to travel down the paper, it is known as *descending technique.*

2. *Ascending Chromatography:* When the development of the paper is done by allowing the solvent to travel up the paper, it is known as ascending technique.

Both ascending and descending techniques have been employed for the separation of biochemical substances. But the descending technique is preferred if the R_F values of various constituents are almost same.

3. *Ascending Descending Chromatography:* it is the hybrid of the above two techniques, In this technique, the upper part of tile ascending chromatography can be folded over a glass rod allowing the ascending development to change over into the descending after crossing the glass rod.

4. *Radial Paper Chromatography:* It is also known as circular paper chromatography. This makes use of radial development. In this technique a circular filter paper is employed. Then the various materials to be analysed are placed at its centre. After drying the spot the paper is fixed horizontally on the petri-dish possessing the solvent so that the tongue or the wick of the paper dips into the solvent. The paper is covered by means of petri-dish cover. The solvent rises through, the tongue or the wick. When solvent front has moved through a sufficient large distance, the components get separate() in the form of concentric circular zones.

5. *Two dimensional chromatography:* In this, a square or rectangular paper is used. The sample is applied to one of the corners. The second development is performed at right angle to the direction of the first run. This type of chromatography can be carried out with identical solvent systems in both the directions or by two solvent systems.

An interesting example of two-dimensional chromatography is the separation of 20 amino acids in the protein hydrolysate. Experimental Details for Quantitative Analysis. For quantitative analysis, the preliminary separation is done in the same way as in qualitative analysis. Then, the assay can be performed either after extraction from the paper or in situ on the paper.

Estimation after extraction from the paper

Isolation of separated components from paper chromatography: The simplest procedure is to cut out the appropriate part of the filter paper having spot and to soak in the minimum quantity of the solvent. Generally, a semi-micio extractor similar to the Soxhelt is being used. The separation of substance from the spot is known as eluation.

Determination: The microanalysis of the resultant eluate can he performed by adopting one or more of the following techniques :

(i) Gravimetric estimation (ii) UV spectrophotometry (iii) Colorimetry (iv) Fluorimetry (v) Polarography (vi) Coulometry (vii) Radioactivity (viii) Flame photometry.

In the assessment of the merits of any procedure, the following information is required.

(i) The nature of the substance to be assayed,

(ii) The scientific equipment available and its sensitivity,

(iii) The time available, and

(iv) The alternative methods, available, if any, and their relative accuracies.

In situ Methods

(i) *By visual assessment: The* simplest procedure is to see the spot by the human eye. However, it is not very quantitative.

(ii) *By measurement of areas:* If the outlines of the spots or zones are well defined, the size of the spot (length or areas) may serve for determinating the quantity of the substance. Then a liner relationship is obtained between the spot area and amount of the substance present,

The random errors in this procedure are generally high as a result of the variations of the spot shape during the separation. Other drawbacks of this method are that the volume of applied sample and the speed of application should be identical in all cases.

(iii) *By Densitometer:* Densitometer is a method whereby the intensity of the colour of a substance is measured directly on the chromatogram. Various workers have used densitometer in the determination of number of organic compounds.

(iv) *Potentiometry:* Changes in the potential of a metal electrode in contact with the filter paper is also utilised with quadrant electrometer of electronic voltmeters.

Besides the above methods, other suitable methods are conductivity, absorptiometry, polarography, etc.

A glass plate is coated, by means of a special device with the material that will constitute the stationary phase mixed with calcium sulphate and made in to a slurry or thin paste sets as a hard surface after drying. The apparatus developed by Stahl for getting stationary phase consists of two parts the aligning tray in which the plates are set in a line, and spreader which takes up the spreading mixture and applies it uniformly on a think over. An improved TLC was developed by DESAGA (West Germany). By this new TLC spreader, the layer thickness can be adjusted from 0.0 to 2.0 mm.

The spreading method is the most important nowadays to obtain thin and uniform layers. The development tank used in paper chromatography is the most common technique in TLC. The bottom of the chamber is covered upto nearly 1 mm by the solvent. Three sides of the tank are lined with solvent impregnated paper while top is covered tightly with the lid as shown. The solvent travels upward by capillarity and in a matter of 1 to 2 hour, it covers the plate. The plate is removed, (fried and otherwise treated just, like" paper.

Most of the methods used for detecting separated solutes on paper are also applicable to TLC. One can see coloured substances usually. However, one can detect colourless components UV light or by treating the plate with a visualising reagent. A large number of available. The corrosive like chromic acid/sulphuric acid can he these he used in TLC.

The possibility of using high temperature techniques such as carbonisation in conjunction with a spary of conc. sulphuric acid offers a universal means of great senstivity. Some interesting applications of thin layer chromatography in biochemistry are given as follows

(i) *Alkaloids: 'TLC* has been used for the isolation and determination of alkaloids in toxicology where the 30-60 minute runs give a great advantage in comparison to the 12-24 hours required for paper chromatography.

Purine alkaloids have been separated by TLC or, silicic acid silica gel and aluminium oxide. The spots are visualised by spraying first with an alcoholic iodine-potassium iodide solution followed by 25% HCI 96% ethanol (1 : 1)

From lobelia syphilitica, pyridine alkaloids were isolated in a pure form by partition and adsorption chromatography. Partition chromatography was carried out on paper, but adsorption chromatography was carried out, but adsorption chromatography was carried out on thin layers of silica gel with chloroform- methanol (3: 1) and on acidic alumina with chloroform ethanol (19 : 1).

(ii) *Amino acids, proteins and Peptides :* Rokkons was successful in isolating 34 ninhydin positive substances isolated from urine and separated on silica gel plates. The chromatograms were developed first with chloroform-methanol-- 17% ammonium hydroxide (2: 2 : 1) and with phenol-water (3 : 1).

Some cyclic peptides has been separated on thin layers of silica gel using a solvent of benzene-ether-methanol (85:10:5).

(iii) *Antibiotics* : Tetracyclines has been separated on thin layers of silica get G by using solvents like 10% citric acid n-butanol methanol 10% citric acid solution (4:2:2) and (4:3:2).

Penicillins have been separated on silica gel 'G' by using the two solvents, acetone- methanol (1:1)) and iso-propanol-methanol (3:7), As the detecting agent, the, iodine-azide reaction was employed by spraying the dried plates with a 0. 1 N iodine solution containing 3.5% of sodium azide. Neomycin sulphates have been separated on layers of activated carbon Both acidic, and neutral layers were used. Active carbon is also used for the separation of water soluble basic antibiotics produced by streptomyces. Erythromycins have been separated on thin layers of silica gel 'G' suing a solvent composed of methyl chloridemethanol-benzene- formamide (80:20:20:2.5).

From the above examples, it can be concluded that the speed, efficiency and sensitivity of thin layer chromatography have made this technique one of the most powerful available to biochemist.

Ion Exchange Chromatography

Ion exchange may be defined as a reversible reaction which free mobile ions of a solid called ion exchange are exchanged for different ions of similar charges present in solution. The principle of ion-exchange chromatography is based upon the simple fact that different cations (or anions) have different capacity to undergo exchange reaction in the surface of a given exchanger.

The capacity of an ion to undergo exchange reaction has been found to depend upon the exchange and lie size of the hydrated ion in solution. Under similar conditions the capacity has been, found to increase with the charge on the ion (i.e., the valency of the ion) but has been found to with the increase in the size of the hydrated ion. Most ion exchangers for practical use consist an, insoluble organic polymer into which a charged group has been some suitable manner.

The backbone is a generally a styrene- divinyl-benzene copolymer. Copolymers of acrylic acid derivatives divinyl-benzene are also frequently used, For use with biological macromolecules, it is usual to introduce charged groups into celluose fibres. For several years, cross-linked dextran (Sephadex) has been also used as carrier material. Cellulose derivatives such as carboxymethyl cellulose (CMC) and diethylaminoethyl cellulose (DEAE) exchangers have been very successfully used in protein purification.

In the ion exchange chromatography, the column is field with any ion exchange which may be either acidic or basic. When a dilutes solution of cation is passed through a column packed with an acid resin, the acid group of resin gets exchanged with this cation. This cation forms complex with resin and it can be recovered by employing an eluting solvent of desired ion strength and pH.

$$\underset{\textit{Resin}}{RH^+} + \underset{\substack{\textit{Cation}\\ \textit{solution}}}{MX^+} \longrightarrow RX + H^+ + M$$

The hydrogen ion released may be washed away with the distilled water. Similarly anion exchange (basic) resins can be employed by using alkalies. When a mixture of substances is put in the column filled with ion exchanger, then by adjusting the pH of the eluting solvent and the ionic strength, the electrostatically held ions are eluted differentially to yield to desired separation.

Applications: The technique of ion exchange chromatography has been widely used for many biochemical mixtures such as proteins nucleic acids, vitamins, sugars, amino acids, etc. Some of these are described as follows:

(i) *Separation of sugars:* This method was developed bp Khym and Zill (1951). These sugar are first of all converted into borate complexes.

Separations of borate complexes of sugars have been achieved on 11 × 0.9 em columns of 200-400 mesh Dowex 1 resin, using a loading of 510 mg of borate complex and flow rates of 0.51 ml/tniri. Quantitative recovery of sugars is possible after separation of the borate complexes. Similarly, disaccharides can he separated from the monsacharides and the individual compounds of hexose and pentose mixtures resolved.

(ii) *Separation of amino acids:* Ion exchange chromatography has been used to separate the complex mixture of 18 amino acids obtained by the acid hydrolysis of proteins. The area under the peaks is a measure of the amount of material used which can he employed for calibration purpose

The mixture of amino acids is first introduced into very short column at pH 2 and eluted with 0.35 N sodium citrate buffer at pH 5.25, with which the column has been equilibrated. Acidic and neutral amino acids at first leave the column unseparated as is shown by the first peaks (M). They are followed by lysine, histidine, NH_3 and arginine. Now a second sample is chromatographed 1 into a longer column with 0.2 N sodium citrate buffer at pH 3.28 and subsequently eluted with 0.88 N citrate buffer at pH 4.25.

(iii) *Purification of Cytochromec:* Cytochromec c has an isoelectric point of pH 10.05. A column containing a cation exchanger buffered at pH 8.5 is prepared. An impure solution of cytochrome at pH 8.5 is placed on the column and water is passed through the column. The contaminating proteins pass freely through the column, (The pI of proteins is usually 7.0 or less) but cytochrome c is held firmly by electrostatic attraction to the resin beads. If the eluting solvent pH is now raised to about 10, the cytochrome c will have a net zero charge and will pass rapidly through as a pure component.

(iv) *Separation of Proteins:* Proteins ar polyionic molecules and like amino acids, most proteins bear a positive charge at acid pH or a negative charge at basic pH. They exchange with one or the other type of cellulose derivative depending on the pH of the solution. They are selectively eluted from the column by buffers of increasing or decreasing pH, For example' if a mixture of proteins is placed on a CMC (carboxymethyl cellulose) at pH 4, the proteins are proteins eluted from the column by increasing the pH of the buffer, Selective elution can also be accomplished by using solutions of salts of increasing ionic strength.

Column Chromatography

Principle: It is known that the rate of adsorption varies with a given adsorbent for different materials. This principle of selective adsorption is used in column chromatography. In this method, the mixture to be separated is dissolved in a suitable solvent and allowed to pass through a tube containing the adsorbent. The component which has greater adsorbing power is adsorbed in the upper part of the column.

The next component is adsorbed in the lower portion of the column which has lesser adsorbing power than the first component. This process is continued. As a result the materials are partially separated and absorbed in the various parts of the column. The initial separation of the various components can be improved by passing either the original or some other suitable solvent slowly through the column. The various bands present in the column become more defined.

The banded column of adsorbent is termed a chromatogram, and the operation is spoken of as the *development of chromatogram.* The portion of a column which is occupied by a particular substance is called its zone. The narrower the zones, the larger the number of substances which can be separated in column of a definite length, and the more concentrated are the eluates. In order to separate or to estimate the various constituents, two procedures may be adopted:

(i) After development, the column of adsorbent maybe pushed out of the tube, the various zones are cut with a knife at boundaries and the substances present in zones extracted with a suitable solvent. This process of recovery of constituents from the chromatogram is known as elution.

(ii) After development the column maybe washed with more solvent, now termed the eluent, and each component is collected separately as it reaches the end of the column and is released.

The essential requirements for adsorption chromatography are a column, an adsorbent and the solvents:

(i) The column is generally a glass tube having means for supporting the packed adsorbent and opening for admission and collection of the mobile phase of the solvent.

(ii) The adsorbent should be inert and insoluble in the solvent used. Sometimes, some inert substances like diatomaceous earth, are mixed with relatively active adsorbent for removing the flow characteristics. The adsorbents such as alumina, $CaCO_2$, (for plant pigments), calcium phosphate (proteins), celite or super cell (proteins) cellulose (organic compounds), glass (high m.w. organic compounds), Kaolin (porphyrins, carbohydrates, vitamin A, D and E), magnesium oxide, magnesium silicate (steroids, carbohydrate derivatives), silicic acid (hydrocarbons, amino acids, carbohydrate derivative) and sucrose (chlorophyll) are commonly used.

(iii) The solvents should be pure and stable. Their recovery from the dissolved substances should be easy. Acetone or organic compounds of this type are generally used in chromatographic separation of plant pigments.

Gas Chromatography

Gas chromatography is quite similar to column chromatography, except that a gas is used as the mobile phase instead of a liquid Gas-solid chromatography (GCS) is based upon selective adsorption on a solid , whereas gas-liquid chromatography is based upon the partition between the gas and an immobile liquid phase.

Instrumentation: The gas chromatographic separation is carried out in a tubular column made of glass, metal or teflon, In this column a sorbent is filled as the stationary. The adsorbents are packed in the form of fine size graded powder, whereas the liquids are either coated as fine film on the column wall or first coated over the liquids graded porous support such as firebrick powder followed by packing in the column. A gas serving as mobile phase through the column. It is known as the carrier gas and serves to transport, sample components in, the column.

The sample is introduced in the vapour form at the

carrier gas entrance end of the column. Different components of the sample are sorbed on the stationary phase to different extents depending upon their distribution coefficients. The portion of each component the gas phase is swept further immediately by the carrier gas. As a result a fraction of the sorbed amount also desorbs out to maintain the k-value. At the same time, out of swept amount some amount will go into sorbent at the next point in the column again to maintain the K-value. This goes on successively and continuously and as a whole the band for each component moves further in the column and having the shape of more or less, Guassian distribution.

It should he kept in mind that the gas is the driving force for the movement of zones through the column and the solid (for GSC or liquid for GLC) provides the selective retarding force. The detector can be regarded as the brain of the chromatograph and the column as its heart. Gas chromatography is an extremely sensitive method capable of detecting traces of substances that can be detected by any other means.

The reason for its sensitivity lies mainly in the detection system at the outlet of the column. There are several types of detectors that measure some property related to the mass of the substances that emerges from the column.

Applications: Gas chromatography has become one of the most widely used chromatographic methods. It is fully automated and extremely versatile but its uses in biochemistry are limited by the instability of many molecules to the high temperatures employed in gas chromatography. It is successfully applied when the mixtures consist of volatile sustances such as fatty acids. The resolving power of a chromatographic column increases with column length and the number of theoretical plates per unit length, although there are limits to the length of a column due to the problem of peak broadening. As the number of theoretical plates is related to the surface area of the stationary phase it follows that the smaller the particle size of the stationary phase, the better the resolution. Unfortunatly, the smaller the particle size, the greater the resistance to eluant flow.

All of the forms of column chromatography so far discussed rely on gravity or low pressure pumping systems for the supply of eluant to the column. The consequence of this is that the flow rates achieved are relatively low and this gives greater time for band broadening by simple diffusion phenomena. The use of faster flow rates is not possible because it creates a back-pressure which is sufficient to damage the matrix structure of the stationary phase, thereby actually reducing eluant flow and impairing resolution. In the past decade there has been a dramatic development in column chromatography technology which has resulted in the availability of new smaller particle size stationary phases which can withstand these pressures and of pumping systems which can c reliable flows rates.

These developments, which have occurred in adsorption, partition., ion-exchange, exclusion and affinity chromatography, have resulted in faster and better resolution and explain why HPLC has emerged as the most popular powerful versatile form of chromatography. Originally, HPLC was referred to as high pressure liquid chromatography but nowadays the term high performance liquid chromatography is preferred since it better describes the characteristics of the chromatography and avoids creating the impression that high pressures are an inevitable prerequisite for high performance. This is now known not to be the case and the term *medium pressure liquid chromatography* (,MPLC) has been coined for some separations. The new technology stationary phases has been applied to thin-layer chromatography giving rise to high *performance thin-layer chromatography* (HPTLC). In general, however, the impact of this new technology has not been quite so great as it has been in column chromatography.

Applications

The wide applicability speed and sensitivity of HPLC has resulted in it becoming the most popular form of chromatography and virtually all type of biological molecules have been purified using the approach. *Reverse phase partition HPLC* is particularly useful for the separation of polar compounds such as drugs and their metabolites, peptides vitamins, polyphenols and steroids. Prior to the advent, of this form of chromatography, the separation of such polar compounds was not easily accomplished and often required pre-derivatisation to less polar *particularly* widely used in clinical.

The technique is particularly widely work as it is possible to apply biological fluids such as serum and urine directly to the column, preferably using a guard column. The separation of some highly polar compounds, such as amino acids, organic acids and the catecholamines which ar difficult to resolve, adequately, by reverse phase chromatography, can often be improved by one of two possible approaches.

The first is ion-suppression in which the ionisation of the compound is suppressed by chromatographing at an appropriately high or low pH. Weak acids, for example, can be chromatographed using an acidified mobile phase. The second is *ion-pairing* in which a *counter ion* with opposite charge to that to be separated is added to the mobile phase so that the resulting ion-pair as sufficient lipophilic character to be retained by the non-polar stationary phase of a reverse phase system. Thus to aid the separation of acidic compounds which would he present as their conjugate anions, a quaternary alkylamine ion such as tetrabutylammonium would be used as the counter ion whereas for the separation of bases which would be present as a cation, an alkyl sulphonate such as sodium heptanesulphonate would be used.

The mechanism by which ion-pairing results in better separation is not clear but two theories have been proposed. The first suggests that the ion-pair behaves as a single neutral species, whilst the second suggests that an active ion-exchange surface is produced in which the counter ion, which has considerable lipophilic properties, and the ions to be separated are adsorbed by the hydrophobic, non-polar stationary phase. In practice, the success of the ion-pairing approach is variable and somewhat empirical.

The size of the counter ion, its concentration and the pH of the solution are all factors which may profoundly influence the outcome of the separation. HPLC has probably had the biggest impact on the separation of oligopeptides and proteins. Instruments dedicated to the separation of proteins have given rise to the technique of *fast protein liquid chromatography (FPLC).* There are no unique principles associated with FPLC, it is simply based on reverse phase and ion-exchange chromatography and on chromatofocussing.

Microbore glass-lined stainless steel columns 1 mm diameter and 2.5 em long have recently been developed which enable very small amounts of sample to be used with separation taking as little as 10 minutes. The technique enables such complex mixtures as tryptic digests of proteins and the culture supernatant of microoragnisms to be applied directly to the column which most commonly contains at, ion-exchange system.

Protein mixtures from cell extracts still need some form of preliminary fractionation prior to study. Although high performance exclusion and ion-exchange chromatography are so successful for protein separation, not all proteins can be completely purified using them. In these cases, the technique of *hydrophobic interaction chromatography which* exploits hydrophobic regions on the surfaces of proteins, may be successful.

The stationary phase is strongly hydrophobic and most commonly is octyl-or phenylagarose. The hydrophobic regions of the proteins' surface interacts with this phase by p p interaction. This minimises interaction of the protein with the aqueous environment. Binding is accomplished in dilute (0.01M) buffer and elution carried out either with aqueous ethylene glycol or ethanol or by the addition of so-called *chaotropic compounds* (perchlorate, trifluroacetate or thiocyanate ions or urea) which disrupt water structure and thus discourage hydrophobic interactions. Proteins purified by this technique include aldolase, transferrin, cytochrame c and thyroglobulin.

Electrophoresis

The powerful method, which was developed by Tiselius in 1937 is based on the principle that there occurs the movement of charged particles under the influence of an electric current to oppositely charged electrodes. Many important biological molecules such as amino acids, peptides, proteins, nucleotides and nucleic acids possess ionisable groups and can therefore be made to exist in solution as electrically charged species, either as cations (+) or anions (–). Even typically non polar substances's such as carbohydrates can be given weak charges by derivatisation, for example as borates or phosphates. Moreover, molecules which have a similar charge will have different charge/mass ratios when they have inherent differences in molecular weight. In combination these differences form sufficient basis for a differential migration when the ions in solution are subjected to an electric field. This is the principle of electrophoresis. The equipment required for electrophoresis consists basically of two items, a power pack and an electrophoresis unit.

Cations move to the cathode and anions move to the anode (+) at rates which depend on the balance between the impelling force of the electric field on the charged ion and the frictional and electrostatic retarding effects between the sample and the surrounding medium. The sample must be dissolved or suspended in buffer for electrophoresis to take place and any supporting medium must also be saturated with buffer to conduct the current. A buffer is also important to maintain a constant state of ionisation since changes in pH would alter the charge on molecules being separated, particularly when sample ions are zwitterions.

The movement of charged particles under the influence on an electric field has been found to depend upon time, electric current, conductivity of the solvent and size and charge of the particle. The distance travelled by the particles in one second under a potential gradient of one volt per centimeter termed as *electrophoretic mobility.* As particles of different sizes are having different electrophoretic nobility, under the influence of some current, it becomes possible to separate them from a mixture. It is also possible to observe electrophoretic movement under the influence of pH gradients also.

Amino acids, for example, do not move to any pole at their isoelectric point because they carry a positive charge at lower pH and move towards cathode and at pH values higher than the isoelectric point they carry a negative charge and move towards anode. The acidic, basic and neutral amino acids will move towards poles at different pH values. It becomes therefore difficult, to maintain a constant pH gradient in the presence of electric current. Experimental. Initially this technique of electrophoresis was used for the separation of proteins in a solution.

The separation by this method, was not quite satisfactory, and several types of supporting materials were used. The charged particles move along this medium, which is generally porous. Most common media are filter paper, cellulose acetate, starch gel, agar and polyacrylamide gel. The filter paper is generally used in routine electrophoretic analyses, although it allows mixing of zones to some extent, due to absorption of molecules on the paper. Commercially available apparatus for electrophoresis are available in two types, the vertical type, in which the supporting medium has been suspended by a glass rod with its end hanging down and the horizontal type in which the medium has been kept horizontal.

The apparatus consists of two electrode tanks which are made to connect though the supporting medium (paper in this case). The paper strip itself has been supported between two glass rods. The electrodes have been connected with a suitable D.C. source. The electrode tanks are filled with a buffer solution. The solution should be selected carefully as some components of the buffer may react with the compounds under separation. The pH of the buffer depends upon the mixture to be separated. Generally, maximum separation has beep attained at the isoelectric point of one of the compounds.

Electrophoresis is mainly used to resolve different components of a protein mixture, Isozymes are also separated by electrophoresis on starch gel or cellulose acetate. An important modification of the electrophoretic technique described above involves the separation of proteins and the other charged molecules in an electric field passing through a solution supported by inert materials such as moistened filter paper, starch gel silica gel, cellulose sponges or glass powder.

This method is known as zone electrophoresis or ionophoresis and is possible to for separation of different mobility into Zones. After separation, it is possible to locate different zones by staining with dyes. Further it is possible to locate different zones individual components from these zones. Thus is a combination of electrophoresis and chromatography. Zone electrophoresis has been found to be useful in the study of serum proteins and the cleavage products of proteins and nucleic acids.

High Voltage Electrophoresis (HVE)

When low molecular weight compounds are separated by low voltage When low molecular weight compounds are paper electrophoresis, considerable diffusion occurs. This can largely be over-come by using much higher voltages, resulting in better resolution and very rapid separations (10 to 60 min), High voltage power packs are available which will supply up to 10 000 V and 500 mA, producing potential gradients of up to 200 supply V cm^{-1}. High voltage electrophoresis generates so much heat that a direct cooling system is required. This is normally achieved by using cooling plates.

The two plates, normally aluminum, are insulated from the supporting medium by polythene and the plates are pressed against the insulated supporting medium by an inflatable pressure pad. Equipment is available with plates of up to 50 x 50 cm, to enable large sheets of paper (generally Whatman 3 MM) to be used. Cold water is circulated through channels in the cooling plates, and for large plates, a flow rate of between 10 and 15 dm^3 min^{-1} is required to disperse the heat produced. In high ambient temperature conditions the cooling water should first be refrigerated; in areas where water is very hard, the use of a water softener is recommended in order to avoid furring up the channels in the plates.

Temperature gradients across the plates must be avoided, since a difference of 1°C causes a 3% change in migrational velocity, and this affects reproducibility. Complete electrical insulation of the equipment is vital because lethal voltages are used. Normal safety precautions should be strictly observed even though commercial systems have a number of cutout (fail safe) devices to avoid accidents. For the separation of protein hydrolysates, HVE in one dimension followed by chromatography in a second dimension is frequently used.

This two-dimensional system allows characteristic fingerprints of different proteins to he obtained, as with two-dimensional thin layer electrophoresis. HVE has proved to be a very effective method for the separation of small peptides and amino acids.

Colorimetry and Spectrophotornetry

Coloured substances tend to absorb certain wave lengths of light and reflect or transmit remaining portion of it. The light reflected or transmitted gives the colour of that substance. The colourless substances have also the property of absorbing certain wavelengths of invisible light, The wavelength at which a particular substance has maxi mum absorption of the light energy is designated by the symbol Max. The values of some important compounds have been given Table 2.

The basic principle underlying absorption of visible or invisible light finds uses in the measurement of the low concentrations of substances. Different types of instruments are available for measuring the absorption of light by coloured or colourless substances. These, are photometers, absorptiometers, colorimeters, spectrometers, and spectrocolorimeters. The principle underlying all these is that the amount of light absorbed by the solution of unknown concentration is compared to that of standard solution and the amount in unknown sample is found out.

Table 2. λ_{max} of some important compounds

Compounds	λ_{max} nm	*Compounds*	λ_{max} nm
Proteins	210 and 280	NAD (P) H	340
Pyrimidines	243	Carotenoids	440
Pyridines	250	Chlorophy ll *a*	663
Nucleotides	260	Chlorophy ll *b*	649
Vitamin A	325	Acetyene	2105-2150

Various types of instruments are available for measuring absorbance. A typical spectrophotometer possesses the following essential components:

(i) Source of light,

(ii) Monochromator unit,

(iii) A detector for light transmitted,

(iv) Meter to read the transmission.

The light source for working in the visible, near infra-red and near ultra violet region is incandescent lamp having tungsten filaments. Hydrogen or deuterium lamps are used for ultra violet regions. In many spectrophotometers, the light is made parallel by using a collimating lens or a mirror before it is passed on to the monochromator unit which is normally made up of a rotating prism and a filter with a slit. Prism resolves light into different wave bands and the filter permits light of certain wave lengths.

The opening of the filter is adjustable. It can be made extremely narrow to obtain resolution upto about 0.5 nm in some spectro-photometers. The instrument has a sample holder which is held between the monochromator unit and the detector sample tunes are made up of glass or fused silica with about 1 min thick wall and 1 em internal diameter. The detector is a photosensitive device which gives a linear response in the spectral region. It has been associated with a electrical device (meter) for reading the exact amount of light passing through the sample.

Calorimetry or spectrophotometry in the visible range has been found to be most useful for determining many inorganic compounds and pigments. Ultra violet spectrophotometry has been used for determining mainly some aromatic and heterocyclic organic compounds. Infra-red spectrophotometry is mainy used for gases. Calorimetry and spectrophotometry are of prime importance in biochemical research. The three important uses are as follows:

(i) If the absorhancy index at a specific wavelength is known, the concentration of a compound can be readily determined, by measuring optical density at that wavelength. As the value of absorbancy index is very large in nucleotides, very small quantities of the absorbing material (2-4mg) can he accurately measured.

(ii) The course of a reaction can be determind by measuring the rate of formation or disappearance of a light absorbing substance, Thus, NADH absorbs strongly at 340 m m whereas the oxidised from NAD^+ has no absorption at this wavelength. Therefore, reactions involving the production, or utilisation of NADH (NADPH) can he assayed by the technique.

(iii) Compounds can be frequently identified by determining the characteristic absorption spectra in the ultraviolet and visible regions of the spectrum.

Radiation Techniques

Radiation techniques are based upon properties of certain isotopic elements which emit radioactive rays such as - α-rays, brays yrays. Like X-rays, these rays could penetrate and affect photographic plates. The elements which emit these rays are called isotopes. Some isotopes disintegrate because particles break off β-particles are electrons;a-particles are helium nuclei (two protons and two neutrons);γ-rays are short wavelength radiation. Such iosotopes are said to be unstable or redioisotopes.

Radioactivity can be detected by devices such as scintillation and Geiger counters. The former are especially useful since they are particularly sensitive to particles emitted by most biologically-important isotopes. particles cause a liquid (scintillator) in the instrument to produce small flashes of light. The light is measured by a meter, which can be scaled to show the amount of radioactivity. Isotopes can be supplied as chemical precursors to trace out biochemical pathways. For example, if 14 CO_2 (radioactive first will presumably be earlier steps in a biochemical pathway than substances which become radioactive later.

This technique is called labelling, and $^{14}CO_2$, is described as a tracer radioisotopes can also label structures. They cause the silver in a photographic film to precipitate as though the film had been made radioactive can thus he identified by the image they form when laid against photographic film. This, technique is called autoradiography. Some isotopes are not radioactive and cannot therefore be detected by Geiger and Scintillation counters. Instead, instruments which separated them on account of their mass, such as mass spectrometers and ultracentrifuges, must, be used. Non-radioactive or stable isotopes have also been in biological research.$H_2^{18}O$, for example, was used to demonstrate that oxygen produced during photosynthesis came from water, not from CO_2, while ^{15}-N-thymidine was used to demonstrate semi - conservative replication of DNA.

Experiment: Radioactive isotopes have been found to be more useful than stable isotopes in understanding the mechanism of reactions as well as processes. In biochemistry, the following two techniques find extensive use in understanding the mechanism of processes.

1. *Tracer technique:* When a biochemist is studying metabolic transformation of a particular substance, his main objective is to observe the fate of that substance in vivo under such experimental conditions which cause minimum physiological disturbance to the test organism. Isotopic tracer technique has physiological disturbance to been found to be most suitable for such type of studies.

In radioactive tracer technique one or more of the atoms in metabolite under study is 'labelled' by means of one of the rare or artificially produced isotopes and its path is followed while frequently testing intermediary compounds at intervals by using sensitive radioactivity detection techniques. By determining precisely the amount of radioactivity in different products the rates and order of their syntheses may be known. This technique is nowadays used for determining the permeability of cells and cell organelles also.

Cells or organelles under testing are exposed for a given period to solutions having radioactive isotopes of sodium or potassium. After sometimes, the cells are centrifuged and the radioactivity in the cells is measured by using a Geiger-Muller counter, By changing the exposure time to solutions having radioactive elements and measuring their levels inside the cell it becomes possible to measure, the rate of uptake of ions. Further, it is possible to measure the rate of loss of ions from the cell by putting these cells in a medium having non-radioactive (cold) element.

The Geiger-Muller Counter: Its working is based upon the ionizing properties of radioactive rays. The line diagram for the various components of this apparatus. In order to measure radioactivity the substance or the cell extract suspected to contain radioactivity is put in the sample tube which is kept in the holder close to the mica window.

Radioactive emission coming from the sample causes ionization of the gas molecules (alcohol vapour + argon) present in the copper tubes of the apparatus. This produces a flash of light in a neon tube and it; is recorded over mechanical register. The conventional G.M. Counter is being replaced by scintillation counter or scintillation spectrometer in the modern biochemistry laboratories. Scintillation counter is somewhat more efficient detector for the low energy radiation.

2. *Autoradiography:* This technique is based upon the ability of the radioactive radiations to blacken a photographic emulsion . After separating a mixture into its components by paper chromatography, the chromatogram i, kept in contact with a photographic film in a dark room. When the film *is* subsequently developed, the radioactive spots on the film appear as blackened spots. The degree of blackening furnishes an idea of the radioactivity present. This technique was used successfully by Melvin Calvin and others in 1950-51 for investgating the pathway of synthesis of carbohydrates from CO_2 in the dark reaction of photosynthesis.

Autoradiography is also being used to determine locations of compounds in a cell or a tissue. For instance, thymidine having 3 H isotope of hydrogen is supplied to actively metabolising cells or tissues for a short duration. The cells are made to count on a slide and fixed. The slide is now kept in contact with a sensitive photographic film. Now this film is developed when we see black spots corresponding to the locations of radioactive materials in the cells. If the black spots are seen over the nucleus, then it can be concluded that thymidine has been incorporated into nuclear contents.

Applications of radioisotopes in the biological sciences

Metabolic Pathways: Radioisotopes are frequently used for tracing metabolic pathways. This usually involves adding a radioactive substrate, taking samples of the experimental material at various times, extracting and chromatographically, or otherwise, separating the products. Radioactivity is then located either by using a Geiger-Muller chromatogram scanner or by obtaining an autoradiogram of the chromatogram. By identifying the labelled compounds, counting the radioactivity in each of them and plotting suitable graphs, it is possible to obtain considerable information on the metabolic pathways involved.

Radioactivity detectors can also be attached to GLC or HPLC chromatographic columns to monitor radioactivity coming off the column during separation. If it is suspected that a particular compound is metabolised by a particular pathway, then radioisotopes can also be used to confirm this. For instance, it Is Possible to predict the fate in individual carbon atoms of ^{14}C-acetate through the tricarboxylic acid cycle (TCA cycle). Methods have been developed whereby intermediates of the cycle can be isolated and the distribution of carbon within each intermediate can be ascertained. This is the so-called *specific labelling pattern.* Should the actual pattern coincide with the theoretical pattern then this is very goods evidence for the operation of the TCA cycle. Another example of the use of radioisotopes to confirm the operation, or otherwise, of a metabolic pathways is in studies carried out on glucose catabolism.

There are numerous ways whereby glucose can he oxidised, the two most important ones in aerobic organisms being glycolysis followed by TCA pathway and the pentose phosphate pathway. Frequently organisms or tissues possess the necessary enzymes for both pathway to occur and it is of interest to establish the relative contribution of each to glucose oxidation.

Both pathways involve the complete oxidation of glucose to carbon (dioxide, but the origin of the carbon dioxide in terms of the six carbon atoms of glucose is different (at least in the initial stages of respiration of exogenously added substrate). Thus it is possible , the carbon dioxide evolved during the respiration of specially labelled glucose (e.g., ^{14}C-6 glucose in which only the C-6 atom is radioactive and n evaluation of the contribution of each pathway to glucose) and obtain a glucose oxidation. The use of radioisotopes in studying the operation of the TCA cycle evaluating the pathway of glucose catabolism are just two examples of how they can be used in confirming metabolic pathways.

Metabolic turnover times: Radioisotopes provide a convenient method of ascertaining turnover times for particular compounds, As an example, the turnover of proteins in rats will be considered. A group of rats is injected with a radioactive amino acid and left for 24 hours, during which time most of the amino acid will be assimilated into proteins. The rats are then sacrificed at suitable time intervals and radioactivity in organs or tissues of interest is determined. In this way it is possible to ascertain the rate of metabolic turnover of protein. Using this sort of method it has been shown that liver protein is turned over in 7 to 14 days, while skin and muscle protein is turned over every 8 to 12 weeks and collagen is turned over at a rate of less than 10% per annum.

Studies of Absorption, Accumulation and Translocation: Radioisotopes have been very widely used in studying the mechanisms and rates of absorption, accumulation and translocation of inorganic and organic compounds by both plants and animals. Such experiments are generally simple to perform and can also yield evidence on the route of translocation and sites of accumulation of molecules of biological interest.

Pharmacological Studies: Another field where radioisotopes are widely used is in the development of new drugs. This is a particularly complicated process since, besides showing if a drug has a desirable effect, much more must be ascertained before it can be used in the treatment of clinical condition. For instance, the site of drug accumulation, the rate of accumulation, the rate of metabolism and the metabolic products must all be determined. In each of these areas of study, radiotracers are extremely useful, if not indispensable. For instance, autoradiography on whole sections of experimental animals yields information on the site and rate of accumulation while typical techniques used, in metabolic studies can be used to follow the rate and products of metabolism

Analytical Applications

Isotope Dilution Analysis: There are many compounds present in living which cannot he accurately assayed by conventional means, since they present in such low amounts and in mixtures of similar compounds Istope dilution analysis offers a convenient and accurate way of overcoming. This problem and avoids the necessity of quantitative isolation. For instance, if the amount of iron in a protein preparation is to be determined, this may he normal methods, but it can be done it a source of ^{59}Fe is available. This is mixed with the protein and a sample of iron is subsequently isolated, assayed for total iron and the radioactivity determined.

If the original specific activity was 10 000 d min^{-1} per 10 mg and the specific activity of the isolated iron was 9000 d min^{-1} 10 mg then the difference is due to the iron in the protein (x). This technique is widely used in, for instance, studies on trace elements.

Radioimmunoassay. One of the most significant advances in biochemical techniques in recent years has been the development of radioimmunoassay (RIA). Radiodating: A quite different analytical use for radioisotopes is in the dating (i.e., determining the age) of rocks, fossil and sediments. In this technique it is assumed that the proportion of an element which is naturally radioactive has been the same throughout time. At the time of fossilisation or deposition the radioactive isotope begins to decay. By determining the amount of radioisotope remaining (or by examining the amount of a decay product) and from a knowledge of the half life, it is possible to date the sample.

For instance, in the radioisotope normally composes 1% of the element and it is found that the sample actually contains 0.25% then two half lives can be assumed to have elapsed since deposition. If the half life is 1 million years then the sample can he dated as being 2 million years old. For long term dating, isotopes with long half lives are necessary, such as ^{235}U, ^{238} U and ^{40}K, whereas for shorter term dating °C is widely used. It cannot he overemphasised that the assumptions made in radiodating are sweeping and hence palaeontologists and anthropologists who use this technique can only give approximate dates to their samples.

Enzyme and Ligand Binding Studies: Virtually any enzyme reaction can be assayed using radiotracer methods as outlined earlier provided that a radioactive form of the substrate is available. Radiotracer-based enzyme assays are more expensive than other methods, but frequently have the advantages of a higher degree of sensitivity. Radioisotopes have also been used in studying the mechanism of enzyme action and in ligand binding studies.

Radioimmunoassay

Radiommunoassay is one of the most important techniques in the clinical and biochemical fields for the quantitative analysis of hormones, steroids and drugs. It combines the specificity of the immune reaction with the sensitivity of radioisotope techniques. Alternatives names used for RIA include saturation analysis, displacement analysis and competitive radioassay. The technique is based on the competitions unlabelled antigen and a finite amount of the corresponding radio- labelled antigen for a limited number of antibody binding sistes in a fixed amount of antiserum.

At equilibrium in antigen excess there will be both free antigen bound to the antibody. Under standard condition the amount of labelled antigen bound antibody will decrease as the amount of unlabelled antigen in the increases. By using known amounts of unlabelled antigen and a fixed amount of antibody and labelled antigen the amount of labelled antigen bound as a function of the total antigen added is measured and a calibration curve constructed. This calibration may then be used to determine the amount, of antigen in sample treated similarly.

The major advantages of RIA include:

(i) the ability to assay any compound that is immunogenic, available in pure form, and can be radio-labelled;

(ii) its high sensitivity some compounds may be detected at a level of pg cm^3

(iii) its high specificity:

(iv) its precision which is comparable to that of other physico-chemical techniques arid far better than bioassays;

(v) its case of automation so that a minimum of manual handling and data processing is necessary. This allows a large number of samples to processed at minimal cost.

The major disadvantages of RIA include:

(i) the relatively high cost of equipment and reagents. Gamma scintillation counters are expensive to buy and maintain, and radioiodine is not a cheap reagent;

(ii) the short shelf-life reagents- the half lives of 125 and 131 are 60 days and 8 days respectively frequent labelling of antisera.

(iii) the radiological hazards of using radioiodine, particularly during the antisera, which must he repeated fairly regulary. Staff should have thyroid scans and be rested if the level of radioactivity increases significantly;

(iv) assays usually take days rather than hours.

Immunoradiometric assay (IRMA)

This utilises purified radio-labelled antibody and is often more sensitive than RIA, for example thyroid stimulating hormone (TSH) may be assayed by allowing it to bind to an antibody complementary to the (x subunit of TSH c attached to Sephadex and then reacting it with radio-labelled antibody against the TSH subunit.

Polarimetry

Introduction: When certain organic liquids, solutions (like sugar) or quartz crystals are placed in the path of plane polarized light, the plane of polarisation is rotated. The property by virtue of which the plane of polarisation of light is rotated is called optical activity and the substances possessing this property arc said to be optically active. Substances which rotate the lane of polarized light toward the right (clockwise) are called dextro rotatory (+) while which rotate towards left (anticlockwise) are called laevo rotatory (-). A mixture of these two varieties in equal proportions will be optically inactive and is called racemic form.

Theory: According to wave theory of light, an ordinary ray of light is considered to he vibrating in all planes at right angles io the direction of its propagation. If this ordinary ray of light is passed through a Nicol prism (a special prism obtained by combining two prisms, of iceland spar suitably cut), the emergent ray has 'its vibrations only in one plane. This light having wave motion in only one plane. This light having wave motion in only one plane is known as plane polarized light.

It has been found that the magnitude of the rotation depends upon the following factors.

(i) Nature of the substance..

(ii) Length of the liquid column (1) through which tight passes.

(iii) Concentration of the solution.

(iv) Nature of the solvent

(v) Temperature of the solution (t).

(vi) Wavelength of the light used.

The rotatory power of a given solution is generally expressed as specific rotation. It is the number of degrees of rotation of the plane polarized light produced by one decimeter in length filled with a solution having one grain of the substance per ml. The measurement is carried out at a temperature t using sodium light. It should be kept in mind that for very dilute solutions in any given non-active solvent, the specific rotation is constant.

Specific rotation be defined as the rotations in degrees of a solution having 1 g material in a 10 cm (1 dm) polarimeter tube. The optical rotation of the compounds has been found to vary according to the wave length of light used, This variation is termed as optical rotatory dispersion.

The apparatus. The light sources is generally the sodium lamp which is emitting light (D line of sodium). The light is made to filter through a slit and rendered which functions as analyser. The entire system is put inside a tube. When there is no optically active sample in polarimeter tube, the lens on analyser side adjusted that the intensity of light seen on the viewer side (eye) becomes minimum. When there is an optically active compound in the tube, the plane of light gets rotated.

The analyser side lens is rotated in order to minimise the light intensity again. This lens is fitted on a scale and the angle of desired rotation can be known from that scale. If we know this absorbed rotation in degrees, tile specific rotation can he calculated as given earlier. 'If the analysing lens is rotated, clockwise, the substance is called dextro-rotatory (+). If the lens is rotated counter -clock wise, the substance is called laevo-rotatory (-)

Uses: Polarimetry is mainly employed for identifying unknown compounds and also for knowing the concentration of known compounds. It can also be employed for studying an enzymatic reaction in which the substrate and product are having different optical rotations. Optical rotatory dispersion measurements can furnish some useful information about the atomic organisation of a molecule. Quantitative methods of analysis by polarimetry are widespread particularly for plant control in the pharmaceutical industry.

The most extensive application, however, is in the sugar industry. In the absence of other optically active substances, sucrose can he determined directly by measuring the angle of rotation, $[\alpha]^{20}_{D}$ being equal to +66.5°. If other active substances are present, it is necessary to measure the optical rotation before and after hydrolysis of the sugar solution.